971174

Study Guide for
Hein, Best, Pattison, and Arena's
Introduction to General, Organic, and Biochemistry
Sixth Edition

Peter Scott
Linn–Benton Community College

Rachael Henriques Porter
University of Illinois, Urbana–Champaign

DISCARDED

Brooks/Cole Publishing Company

I(T)P® An International Thomson Publishing Company

Pacific Grove • Albany • Belmont • Bonn • Boston • Cincinnati • Detroit
Johannesburg • London • Madrid • Melbourne • Mexico City
New York • Paris • Singapore • Tokyo • Toronto • Washington

Assistant Editor: *Beth Wilbur*
Cover Design: *Vernon T. Boes*
Cover Photo: *Photonica International*
Editorial Associate: *Nancy Conti*
Marketing Team: *Kathleen Sharp, Christine Davis, Carrie Beckwith*
Production Editor: *Mary Vezilich*
Printing and Binding: *Patterson Printing*

COPYRIGHT© 1997 by Brooks/Cole Publishing Company
A division of International Thomson Publishing Inc.

 The ITP logo is a registered trademark under license.

For more information, contact:

BROOKS/COLE PUBLISHING COMPANY
511 Forest Lodge Road
Pacific Grove, CA 93950
USA

International Thomson Publishing Europe
Berkshire House 168-173
High Holborn
London WC1V 7AA
England

Thomas Nelson Australia
102 Dodds Street
South Melbourne, 3205
Victoria, Australia

Nelson Canada
1120 Birchmount Road
Scarborough, Ontario
Canada M1K 5G4

International Thomson Editores
Seneca 53
Col. Polanco
11560 México, D. F., México

International Thomson Publishing Japan
Hirakawacho Kyowa Building, 3F
2-2-1 Hirakawacho
Chiyoda-ku, Tokyo 102
Japan

International Thomson Publishing Asia
221 Henderson Road
#05-10 Henderson Building
Singapore 0315

International Thomson Publishing GmbH
Königswinterer Strasse 418
53227 Bonn
Germany

All rights reserved. No part of this work may be reproduced, stored in a retrieval system, or transcribed, in any form or by any means—electronic, mechanical, photocopying, recording, or otherwise—without the prior written permission of the publisher, Brooks/Cole Publishing Company, Pacific Grove, California 93950.

Printed in the United States of America

5 4 3 2 1

ISBN 0-534-25879-4

Dedicated to my dear wife, Jeanette

- P.S.

Dedicated to my parents

- R.H.P.

Preface

This study guide has been prepared to accompany *Introduction to General, Organic and Biochemistry, Sixth Edition*, by Morris Hein, Leo Best, Scott Pattison and Susan Arena. The guide selects certain key concepts from the text, and provides the student with a means of self-evaluation for determining how well he or she understands the material of each chapter.

By presenting slightly different viewpoint and emphasis, the study guide provides students with an approach other than that of the textbook toward mastery of the subject matter. Because this guide is an auxiliary student-oriented aid to complement the textbook, no new material is presented.

The chapters in the guide are organized as follows: The chapter headings are the same as in the text with a self-evaluation section providing various exercises that allow the student to check his or her understanding of the text chapter material. Students are asked to complete fill-in responses or to choose an appropriate response. Tables to be constructed and problems to be solved are presented. Nomenclature, equation balancing, and use of the periodic table are covered. The last problems in the self-evaluation section are identified as challenge problems. These will usually be a little more difficult or complex than the other exercises in the chapter. Additionally, the explanations about the answer will be briefer and will assume that the student has a good grasp of the basic material covered up to that particular point. Complete answers to questions and solutions to problems are given at the end of each chapter, to provide immediate reinforcement or correction of errors. The last section in each chapter provides a recap or summary.

In addition to the usual exercises, several puzzles are included in the guide to give the student a break and to check his or her vocabulary skills. The student will probably need a periodic table to help in solving the puzzles.

The author visualizes using the study guide as follows. The student would (1) read the assigned text chapter; (2) go through the self-evaluation process in the guide; (3) return to the assigned homework problems, lab experiments, and other instructor-generated activities; and (4) be tested on predetermined performance objectives for each chapter.

To the Student

This guide, which accompanies the textbook *Introduction to General, Organic and Biochemistry, Sixth Edition*, by Morris Hein, Leo Best, Scott Pattison and Susan Arena is a student-oriented self-study guide. It has been prepared to help you evaluate your grasp of certain concepts in the textbook; you can then identify trouble spots and ask your instructor for individual help.

What is the place of the study guide in the total chemistry program? Your instructor will hand out reading assignments from the textbook, which will be followed by homework questions and problems and laboratory experiments. After classroom discussion and problem sessions, you will be tested on the material. So, when do you use the study guide? A typical student of mine might first read the assigned chapter in the text (with rereading, if necessary), and then turn to the study guide for a self-evaluation of his or her understanding of the key concepts. All of the answers and solutions are given so that any errors can be quickly spotted. If a particular concept or type of problem doesn't make sense, the student can return to the textbook. Then the student is ready to work the homework assignment, knowing that he or she understands the basic concepts in the chapter. The week's lab experiment is performed, followed by the chapter quiz.

Each chapter of the study guide that has questions to answer and problems to solve is organized as follows: The "Self-Evaluation Section" provides various types of exercises that enable you to check your understanding of the key objectives. The answers and solutions to the problems are given at the end of the chapter so that you can find out where your difficulty may have occurred. The challenge problems at the end of each chapter are usually a bit more difficult or complex than the rest of the self-evaluation exercises. You should work on these after you feel comfortable with the other material. The answers to the challenge problems may not be as thorough as the other answers since you're expected to have a good grasp of the basic material before attempting the challenge problems. The last section gives a summary from the chapter and indicates where you are going next in the course.

The study guide allows you to work on your own and make efficient use of your time. It should assist you in finding out whether you have learned the basic concepts.

Best wishes for success in your study of chemistry.

CONTENTS

Chapter	1	An Introduction to Chemistry	1
Chapter	2	Standards for Measurement	7
Chapter	3	Classification of Matter	19
Chapter	4	Properties of Matter	27
Chapter	5	Early Atomic Theory and Structure	35
Chapter	6	Nomenclature of Inorganic Compounds	41
Chapter	7	Quantitative Composition of Compounds	47
Chapter	8	Chemical Equations	61
Chapter	9	Calculations from Chemical Equations	69
Chapter	10	Modern Atomic Theory	79
Chapter	11	Chemical Bonds: The Formation of Compounds from Atoms	87
Chapter	12	The Gaseous State of Matter	97
Chapter	13	Water and the Properties of Liquids	111
Chapter	14	Solutions	119
Chapter	15	Acids, Bases, and Salts	145
Chapter	16	Chemical Equilibrium	157
Chapter	17	Oxidation-Reduction	177
Chapter	18	Nuclear Chemistry	191
Chapter	19	Chemistry of Selected Elements	197

Chapter	20	Organic Chemistry: Saturated Hydrocarbons	206
Chapter	21	Unsaturated Hydrocarbons	213
Chapter	22	Alcohols, Ethers, Phenols, and Thiols	220
Chapter	23	Aldehydes and Ketones	227
Chapter	24	Carboxylic Acids and Esters	234
Chapter	25	Amides and Amines: Organic Nitrogen Compounds	243
Chapter	26	Polymers: Macromolecules	250
Chapter	27	Stereoisomerism	257
Chapter	28	Carbohydrates	266
Chapter	29	Lipids	277
Chapter	30	Amino Acids, Polypeptides, and Proteins	282
Chapter	31	Enzymes	288
Chapter	32	Nucleic Acids and Heredity	292
Chapter	33	Nutrition	298
Chapter	34	Bioenergetics	303
Chapter	35	Carbohydrate Metabolism	308
Chapter	36	Metabolism of Lipids and Proteins	313

CHAPTER ONE

INTRODUCTION

SELF-EVALUATION SECTION

We will begin your use of the study guide with some straight-forward true-false and fill-in exercises followed by a quick mathematics survey section. As mentioned in the introduction, the use of mathematics is very important for your success in chemistry.

1. Decide if the following statements are True or False:

    (1) The field of chemistry is not related to the applied sciences such as agriculture, oceanography, medicine and fire science.

    (2) Leaning about the risks and benefits associated with chemicals will help you be a better informed citizen.

    (3) Studying chemistry will not help you improve your problem-solving techniques.

    (4) Chemistry is the science of the composition, structure, properties and reactions of matter.

    (5) Living organisms can be thought of as a system of interrelated chemical processes.

    (6) Chemists work alone in isolated laboratories in order to undersrtand the nature of chemical reactions.

2. The scientific method is a general approach to answer questions or solve problems. Each of the following statements is missing an "action verb" which is to be selected from the list below.

    Use these verbs to fill in the blanks:

    formulate
    modify
    analyze
    plan and do
    collect

(1) _____ the facts or data that are relevant to the question which is usually done by experimentation.

(2) _____ the data to find trends.

(3) _____ a hypothesis that will account for the accumulated data and can be tested by further experimentation.

(4) _____ additional experiments to test the hypothesis.

(5) _____ the hypothesis as necessary so that it is compatible with all pertinent data.

3. Decide if the following statements are True or False.

(1) A well-established hypothesis is called a scientific law.

(2) Chemistry is an experimental science.

(3) To be successful as a chemist today requires teamwork and cooperation.

(4) Risk assessment of chemical hazards is solely the work of chemists.

(5) In general, any chemical that is "synthetic" is bad and anything "organic" is good.

(6) Risk management invloves value judgments.

(7) Risk assessment is a process that attempts to determine the risk associated with exposure to a certain chemical.

(8) In some cases, all risks associated with exposure to certain chemicals can be eliminated.

4. As a good way to check your readiness to begin a course in chemistry, work through the following mathematics survey. We use this survey for all of our chemistry students, and they have found it helpful in determining areas to be reviewed.

Circle the response for each of the questions that most closely expresses your feeling, as follows:

(a) The problem is worked correctly.
(b) The problem has been worked incorrectly.
(c) I have studied this type of problem before but can't remember how to work it now. I'm not sure if it's right or not.
(d) I have never studied this type of problem before.

(1) $2/4 = 0.33$ a b c d
(2) $1/16 = 0.0625$ a b c d
(3) $1/10 = 0.01$ a b c d
(4) $30/6 = 0.5$ a b c d
(5) $1/3 \times 4/5 = 5/8$ a b c d
(6) $1/2 \div 1/4 = 2$ a b c d
(7) $3/4 + 3/5 = 3/20$ a b c d
(8) $1/3 - 1/4 = 1/12$ a b c d
(9) $\dfrac{2 \times 4 \times 3}{6 \times 2} = 2$ a b c d

(10) $10^2 = 100$ a b c d
(11) $\sqrt{25} = 5$ a b c d
(12) $2^3 = 6$ a b c d
(13) $\sqrt[3]{8} = 2$ a b c d
(14) $6^2 \times 6^3 = 6^6$ a b c d

(15) $\dfrac{9^4}{9^3} = 9$ a b c d

(16) $\dfrac{a^7}{a^{10}} = a^{-3}$ a b c d

(17) $6.9 \times 10^3 = 690$ a b c d
(18) $0.0054 = 5.4 \times 10^{-3}$ a b c d
(19) $(2 \times 10^3)(5 \times 10^2) = 10^6$ a b c d
(20) $(5 \times 10^{-3})(3 \times 10^5) = 1.5 \times 10^2$ a b c d
(21) 1594.61 rounded off to four figures is 1596   a b c d
(22) The reciprocal of 2 is 1/2   a b c d
(23)   6.1 cm
     $\times$ 3.5 cm
       305
       183
     21.35 cm^2 Is this answer written with the correct number of significant figures?   a b c d

(24) If $\dfrac{x}{2.5} = \dfrac{3}{7.5}$, then $x = 1$ a b c d

(25) Let $T_1 \times N_1 = T_2 \times N_2$
If $T_1 = 5$, $N_1 = 3$ and $T_2 = 6$,
then $N_2 = 2.5$ a b c d

(26) If $\dfrac{P_1 V_1}{T_1} = \dfrac{P_2 V_2}{T_2}$, then $T_2 = \dfrac{P_2 V_2}{P_1 V_1 T_1}$ a b c d

(27) If $\dfrac{1.5}{x} = \dfrac{3.0}{6.4}$, then $x = 3.5$ a b c d

(28) If it takes 1 calorie to heat 1 gram of water 1°C, how many carories will be required to heat 3 grams of water 5°C? Answer: 8 calories   a   b   c   d

Check your answers from the list on page 5. Students who miss as many as 17 will have a difficult time with the material in the text without an intensive review and extra work outside of class. Keep this in mind if you found that many of the problems were unfamiliar to you or you circles an incorrect response. If you missed about half of the problems, you should study thoroughly Appendix I — Mathematical Review at the back of the text.

RECAP SECTION

The objectives of Chapter 1 are to introduce you to the broad field of science known as chemistry. We discuss the importance of chemistry in other fields of science such as biology and agriculture.

You should also gain some insight into how the scientific method is applied in chemistry. As you study chemistry or any experimental science, you find the steps of hypothesis, theory, and scientific law leading time and again toward an understanding of natural phenomena.

The importance of problem solving to the science of chemistry cannot be over-emphasized. You will be working problems within the text material, solving homework problems, and using this study guide to evaluate your ability to solve certain specific types of problems. Many of the laboratory experiments also involve problem solving.

The techniques you will use are mostly simple arithmetic operations, but you must have a complete understanding of them. Significant figures and scientific notation are also important for your progress in chemistry.

The topics of scientific notation, significant figures and rounding-off numbers are covered thoroughly in Chapter 2 of the text. Remember that additional helpful information which will be of use throughout your study of chemistry is found in Appendix I of the text.

ANSWERS TO QUESTIONS AND MATHEMATICS SURVEY

1. (1) False   (2) True   (3) False   (4) True   (5) True   (6) False

2. (1) collect   (2) analyze   (3) formulate   (4) plan and do   (5) modify

3. (1) False   (2) True   (3) True   (4) False   (5) False   (6) True
   (7) True   (8) False

4. (1) b   (8) a   (15) a   (22) a
   (2) a   (9) a   (16) a   (23) b
   (3) b   (10) a   (17) b   (24) a
   (4) b   (11) a   (18) a   (25) a
   (5) b   (12) b   (19) a   (26) b
   (6) a   (13) a   (20) b   (27) b
   (7) b   (14) b   (21) a   (28) b

CHAPTER TWO

Standards for Measurement

SELF-EVALUATION SECTION

If you are to talk correctly with your fellow students and instructors about chemical principles and knowledge, you must have an intimate understanding of chemical terminology. Chapter 2 contains some very essential terms, which you must master. Some of the terms will be familiar but will have more precise definitions than you are used to. For instance, though the terms *mass* and *weight* are often used interchangeably, the two words do not have precisely the same meaning. Mass is an inherent property of matter and constitutes an essential characteristic of matter. Weight, on the other hand, is an extraneous property of an object and depends on the force of gravity acting on an object. An object that weighs 100 grams at sea level will weigh less at an altitude of 10,000 feet since the force of gravity is less. If the same object is placed in a space vehicle, it will exhibit weightlessness at some point in space and float around the cabin if not anchored down.

However, if a chemical balance, which is used to measure masses of objects by comparison with known masses, is used to make a mass determination of an object and a value of 225 grams is obtained, then the object's mass is constant no matter where the object is. A determination of the object's mass in a space vehicle would yield the same value, 225 grams, as before. If two objects of equal mass are weighed at the same place on the earth's surface they will have the same weight. The unfortunate circumstance is that there is no simple word in English to describe the operation of making a mass determination, so the verb "to weigh" is used for this important experimental procedure. Keep in mind that when you make a weighing in the chemistry laboratory you are in fact measuring the mass of the object being weighed.

Two other terms introduced in Chapter 2 that are used in everyday conversation are *heat* and *temperature*. There is some confusion about the proper usage of *temperature* and *heat*. It is usual to state that the temperature of an object indicates how much heat the object contains. But this is an incorrect notion. Think of two objects of different size but at the same temperature. Which object contains more heat energy? The larger object, of course. Thus, temperature is only an indication of the degree of "hotness" or "coldness" of an object. The temperature scales are relative, arbitrary scales established with certain reference points. The Celsius scale, for instance, has 0 degrees set at the freezing point of water and 100 degrees set at the boiling point of water.

Heat is one form of energy just as light and electricity are. When small particles are vibrating, heat energy is given off. Think of the conversion of electrical energy into heat and light energy as an electric stove burner becomes red-hot. It is not too difficult to imagine the intense motion in the heating element as the current of electricity passes through.

The last term defined in this chapter is *density*. Again, we can use either an algebraic form of definition or the English equivalent. The density of a substance is related to its mass and volume. Since all matter in the universe has mass and occupies space or volume, the quantity defined as density is a useful physical characteristic used in describing various substances. The defining equation for density says that the density of a substance is equal to the mass divided by the volume. The mass is given in grams, and the volume can be expressed in cubic centimeters, millimeters, or liters. The units for density are therefore either grams/cubic centimeter (g/cm^3), grams/ milliliter (g/mL), or grams/liter (g/L).

We have just finished discussing a few of the important new terms used in chemistry. Did you notice that the definition of *density* involved a simple algebraic equation with units? Thus, density is defined as the mass per unit volume ($d = m/V$, for example). No numbers or measurements are involved in these definitions. After completing a review of the metric system, you will work through sample problems involving the densities of substances using the factor-label or dimensional analysis technique. The factor-label approach will be used as consistently as possible throughout the study guide.

1. In order to handle large numbers efficiently, it is worthwhile to use scientific notation, which means writing a number as a power of 10. The numbers are always written between 1 and 10 with the associated power of 10.

    (1) Express the following numbers as powers of 10:

    2510   _____

    0.0065   _____

    46   _____

    77,000,000   _____

    0.0102   _____

    (2) Express the following in decimal form:

    4.77×10^4 _____

    8.41×10^{-2} _____

    5.8×10^1 _____

    9.1×10^0 _____

    1.415×10^2 _____

2. Review of metric units, abbreviations, and definitions. Fill in the correct responses.

    In our study of chemistry, we will be using the metric system of measurement routinely. On any one occasion, we may have the task of making measurements of one of several quantities, which might include (1) _____, (2) _____, or

(3) _____. For any measurement of mass in the metric system, we will use the unit name (4) _____ with the abbreviation (5)_____. If we are measuring the length of an object, as in the case of certain density experiments, we will use the unit called (6)_____, which is abbreviated (7) _____. However, probably the most common type of measurement we make in a chemistry laboratory involves the metric volume (8) _____, which is abbreviated (9) _____.

To these three units of measure for mass, length, and volume, we can attach prefixes to change the unit's value by various powers of 10. For example, *kilo* means (10) _____ and *milli* means (11) _____. The prefix for 0.01 (10^{-2}) is (12) _____, and the prefix for 0.000001 (10^{-6}) is (13) _____. The abbreviations for the pre-fixes *kilo, centi, micro,* and *milli* are (14) _____, (15) _____, (16) _____, and (17) _____.

Using the four prefixes in general use in chemistry, we must be able to convert from one metric unit to any other corresponding metric unit. We will use a series of conversions to find out if a series of decimal-point changes presents any problems to you.

Convert These Units   **To These Units**

(18) 42 g   _____ mg

(19) 0.741 cm   _____ m

(20) 8.4 mL   _____ L

(21) 776 kg   _____ g

(22) 1005 µL   _____ mL

(23) 15 µg   _____ mg

(24) 6.2 L   _____ mL

(25) 0.34 km   _____ cm

Convert These Units   **To These Units**

(26) 1.25 mg   _____ g

(27) 0.0089 g   _____ mg

(28) 9 cm   _____ µm

3. Express the following numbers in exponential form:

   (1) 67,000 _____

   (2) 0.0654 _____

   (3) 10,000,000 _____

   (4) 0.00078 _____

   (5) 411,000 _____

   (6) 10 _____

4. Express the following numbers in decimal form:

   (1) 4.8×10^3 _____

   (2) 0.67×10^2 _____

   (3) 0.151×10^{-5} _____

   (4) 1074×10^{-2} _____

   (5) 0.0034×10^3 _____

   (6) 11×10^{-5} _____

5. Round off the following numbers to four significant digits:

   (1) 47.679 _____

   (2) 1.0255 _____

   (3) 100.2484 _____

   (4) 16.0502 _____

   (5) 0.077938 _____

   (6) 21.6251 _____

   (7) 5.00515 _____

   (8) 74.2856 _____

6. Solve for x.

   (1) $x = (1.3 \times 10^2)(6.4 \times 10^{-1})$

   (2) $x = (5.76 \times 10^{-4})(2.4 \times 10^3)$

   (3) $(8.9 \times 10^1) x = (5.46 \times 10^3)$

   (4) $(0.77 \times 10^4) x = 125{,}000$

7. Converting American units to metric units.

   (1) A college basketball player playing forward is generally about 6 ft 6 in. tall. How many centimeters is this? How many meters?

   Do your calculations here.

   (2) Tall cans of juice contain 46.0 fluid ounces. How many mL is this? One ounce equals 29.6 mL.

   Do your calculations here.

(3) A can of green beans costs 59¢ and weighs 1 pound. What is the cost in cents per gram? The words "cents per gram" are a way of expressing the fraction cents/gram. This means that the number of cents is divided by the number of grams. The word "per" is a common method used in science to indicate a division operation or a fraction. Be alert to this term in future problems.

$$1 \text{ lb} = 453.6 \text{ g}$$

Do your calculations here.

8. We can convert the Kelvin temperature scale to the Celsius scale, and vice versa, with the formula $K = °C + 273.15$ The value 273.15 is a constant value. Using the formula, make the following conversions.

Convert From   **To**

(1) 25°C   _____ K

(2) 273K   _____ °C

(3) –15°C   _____ K

(4) –273°C   _____ K

(5) 398K   _____ °C

(6) 100°C   _____ K

(7) 58K   _____ °C

Using the equations given for converting °F to °C, and vice versa, make the con-versions asked for.

$$°F = (1.8 \times °C) + 32$$

$$°C = \frac{(°F - 32)}{1.8}$$

(8) It is not uncommon for temperatures in the Canadian plains to reach –60°F and below during the winter. What is this temperature in °C?

Do your calculations here.

(9) Sodium chloride melts at 801°C. What is this temperature in °F?

Do your calculations here.

9. As you progress in your study of various substances, you will find a great diversity in the physical nature of matter. There are many ways to categorize matter in trying to relate the properties of one substance to another. One such way is through a relationship called *density*. The density of a substance is not only related to its mass but also to the volume it occupies. Thus, a cube of aluminum metal measuring 10 cm on a side has much less mass than a cube of iron metal of the same volume. We say the density of aluminum is less than that of iron.

The equation defining density is written as follows:

$$\text{Density} = \frac{\text{mass}}{\text{volume}}$$

$$d = \frac{g}{cm^3} = \frac{g}{mL} \text{ or } \frac{g}{liter}$$

The units for density vary depending on the units used for volume.

(1) Sample problem: Our block of aluminum, 10 cm on a side, was found to have a mass of 2700. g by using a balance. What is the density of aluminum?

Do your calculations here.

(2) Another problem: Logs of black ironwood, which do not float in water, are found to have a density of 1.077 g/cm^3. What is the volume of a piece of ironwood that has a mass of 750. g?

Do your calculations here.

Challenge Problems

10. The standard of metric length is the meter which was defined (prior to 1983) as 1,650,763.73 wavelengths of a spectral line of the element krypton. To three significant figures determine this wavelength in nanometers if $1 nm = 10^{-9}$ m.

    Do your calculations here.

11. For each of the following, identify (a) what is the numerical value and what is the unit, (b) how many significant figures, (c) which numbers are known and which are estimated. Round each number to two significant figures (d), and (e), express that number as a power of 10.

    (1) 1055 kg

    (2) 1,650,763 wavelengths

    (3) 1.077 g/cm^3

    (4) 0.00845 L

    (5) 776,000 g

RECAP SECTION

This has been a long chapter, but if you feel comfortable with the new terms and techniques, you've come a long way toward being successful in the remaining weeks of the course. We have reviewed some basic terms and concepts, brushed up on the metric system, and practiced converting one metric unit to another. The temperature scales of the most concern are the Celsius and Kelvin scales. The conversion between the two scales is very simple since the size of the degree is the same. Perhaps the most important aspect of the chapter was putting your algebraic skills to use in various practical ways to solve chemical problems. You will find that most chemical problems can be handled by a few simple algebraic techniques. If you were successful with the problems in Chapter 2, you are ready to proceed. If you had difficulty, you should go back over the Math Review in the text and then return to the study guide.

ANSWERS TO QUESTIONS AND SOLUTIONS TO PROBLEMS

1. (1) 2.51×10^3; 6.5×10^{-3}; 4.6×10^1; 7.7×10^7; 1.02×10^{-2}
   (2) 47,700; 0.0841; 58; 9.1; 141.5

2. (1) (2) (3) mass, length, volume, density, or temperature are possible answers
   (4) kilogram (5) kg (6) meter (7) m (8) liter
   (9) L (10) 1000 (11) 0.001 (12) centi- (13) micro-
   (14) k (15) c (16) μ (17) m

   (18) 4.2×10^4 mg $\quad 42 \text{ g} \times 1000 \dfrac{\text{mg}}{1 \text{ g}} = 42{,}000 \text{ mg} = 4.2 \times 10^4 \text{ mg}$

   (19) 0.00741 m $\quad 0.741 \text{ cm} \times \dfrac{1 \text{ m}}{100 \text{ cm}} = 0.00741 \text{ m}$

   (20) 0.0084 L $\quad 8.4 \text{ mL} \times \dfrac{1 \text{ L}}{1000 \text{ mL}} = 0.0084 \text{ L}$

   (21) 7.76×10^5 g $\quad 776 \text{ kg} \times \dfrac{1000 \text{ g}}{1 \text{ kg}} = 776{,}000 \text{ g} = 7.76 \times 10^5 \text{ g}$

   (22) 1.005 mL $\quad 1005 \text{ } \mu\text{L} \times \dfrac{1 \text{ mL}}{1000 \text{ } \mu\text{L}} = 1.005 \text{ mL}$

   (23) 0.015 mg $\quad 15 \text{ } \mu\text{g} \times \dfrac{1 \text{ mg}}{1000 \text{ } \mu\text{g}} = 0.015$

   (24) 6.2×10^3 mL $\quad 6.2 \text{ liter} \times \dfrac{1000 \text{ mL}}{1 \text{ liter}} = 6200 \text{ mL} = 6.2 \times 10^3 \text{ mL}$

   (25) 3.4×10^4 cm $\quad 0.34 \text{ km} \times \dfrac{1000 \text{ m}}{1 \text{ km}} \times \dfrac{100 \text{ cm}}{1 \text{ m}} = 34{,}000 \text{ cm} = 3.4 \times 10^4 \text{ cm}$

   (26) 0.00125 g $\quad 1.25 \text{ mg} \times \dfrac{1 \text{ g}}{1000 \text{ mg}} = 0.00125 \text{ g}$

   (27) 8.9 mg $\quad 0.0089 \text{ g} \times \dfrac{1000 \text{ mg}}{1 \text{ g}} = 8.9 \text{ mg}$

   (28) 9×10^4 μm $\quad 9 \text{ cm} \times \dfrac{1 \text{ m}}{100 \text{ cm}} \times \dfrac{10^6 \text{ } \mu\text{m}}{\text{m}} = 9 \times 10^4 \text{ } \mu\text{m}$

3. (1) 6.7×10^4 (4) 7.8×10^{-4}
   (2) 6.54×10^{-2} (5) 4.11×10^5
   (3) 1×10^7 (6) 1×10^1

4. (1) 4800 (4) 10.74
   (2) 67 (5) 3.4
   (3) 0.00000151 (6) 0.00011

5. (1) 47.68 (4) 16.05 (7) 5.005
   (2) 1.026 (5) 0.07794 (8) 74.29
   (3) 100.2 (6) 21.63

6. (1) $x = 83$
   (2) $x = 1.4$
   (3) $x = 61$
   (4) $x = 16$

7. (1) We need to change 6 ft 6 in. into inches and then convert to the metric units of centimeters and meters.

   $$6 \text{ ft } 6 \text{ in.} = 6 \text{ ft} \times \frac{12 \text{ in.}}{\text{ft}} + 6 \text{ in.} = 78 \text{ in.}$$

   $$78 \text{ in.} \times \frac{2.54 \text{ cm}}{\text{in.}} = 2.0 \times 10^2 \text{ cm}$$

   $$2.0 \times 10^2 \text{ cm} \times \frac{1 \text{ m}}{100 \text{ cm}} = 2.0 \text{ m (2 significant figures)}$$

   (2) Since 1 oz. = 29.6 mL, we need to set up the problem to cancel out oz. and give an answer with units of mL.

   $$46.0 \text{ oz.} \times \frac{29.6 \text{ mL}}{1 \text{ oz.}} = 1.36 \times 10^3 \text{ mL (3 significant figures)}$$

   (3) 13¢ per g
   We must first convert 1 pound to grams and then divide the cost by the number of grams.

   $$1 \text{ lb} \times \frac{453.6 \text{ g}}{\text{lb}} = 453.6 \text{ g}$$

   59¢/454 g = 13¢ per g (2 significant figures)

8. (1) 298K (2) 0°C (3) 258K (4) 0K
   (5) 125°C (6) 373K (7) −215°C

   (8) $°C = \frac{-60° - 32}{1.8} = \frac{-92°}{1.8} = -51°C$

   (9) $°F = (1.8 \times 801°) + 32 = 1470°F$

— 15 —

9. (1) 2.7 g/cm^3

The mass is given as 2700 g, and we need to determine the volume of a cube 10 cm on a side. Volume is length × width × height. Multiplying 10 cm × 10 cm × 10 cm equals 1000 cm^3. Substituting into our equation, we have

$$d = \frac{\text{mass}}{\text{volume}} = \frac{2700 \text{ g}}{1000 \text{ cm}^3} = 2.7 \frac{\text{g}}{\text{cm}^3}$$

The answer says that aluminum metal has a mass of 2.7 g per each cm^3. Therefore, 2 cm^3 would have a mass of 5.4 g, and so forth.

(2) 696 cm^3

We are given the density of the wood and the mass and asked to solve for volume. We write down the equation, then isolate the unknown quantity and take a close look at how the units of a problem can help us in solving equations.

$$d = \frac{\text{mass}}{\text{volume}}$$

Multiply each side by volume.

$$\text{volume} \times d = \frac{\text{mass} \times \cancel{\text{volume}}}{\cancel{\text{volume}}}$$

Now divide each side by density.

$$\frac{\text{volume} \times \cancel{d}}{\cancel{d}} = \frac{\text{mass}}{d}$$

Substitute our values into the modified equation.

$$\text{volume} = 750.\cancel{\text{g}} \times \frac{1 \text{ cm}^3}{1.077 \cancel{\text{g}}}$$

$$= 696 \text{ cm}^3 \text{ (3 significant figures)}$$

Notice that "g" cancels out then leaves our answer with the units for volume, which are the desired units.

10. From the information given, if $1 \text{ nm} = 10^{-9} \text{ m}$, then $1 \text{ m} = 10^9 \text{ nm}$. Therefore, we can say that

$$1 \text{ m} = 10^9 \text{ nm} = 1{,}650{,}763.73 \text{ wavelengths}$$

To find the length of one wavelength we can set up the following two relationships:
$$10^9 \text{ nm} = 1{,}659{,}763.73 \text{ wavelengths}$$

$$\frac{10^9 \text{ nm}}{1{,}650{,}763.73} = 1 \text{ wavelength (dividing each side by 1,650,763.73)}$$

$$606 \text{ nm} = 1 \text{ wavelength (to three significant figures)}$$

11. (1) 1055 kg: (a) 1055 is the numerical value, kg is the unit (b) 4 significant figures (c) 105 are known, last 5 estimated (d) 1100 kg to 2 significant figures (e) 1.1×10^3 kg

(2) 1,650,763 wavelengths: (a) 1,650,763 is the numerical value, wavelengths are the units (b) 7 significant figures (c) 165076 are known, last 3 is estimated (d) 1,700,000 wavelengths to 2 significant figures (e) 1.7×10^6 wavelengths

(3) 1.077 g/cm^3: (a) 1.077 is the numerical value, g/cm^3 are the units (b) 4 sigificant figures (c) 1.07 known, last 7 estimated (d) 1.100 g/cm^3 to 2 significant figures (e) 1.1×10^0 g/cm^3

(4) 0.00845 L (a) 0.00845 is the numerical value, L is the unit (b) 3 significant figures (c) 84 known, 5 estimated (d) 0.0085 L to 2 significant figures (e) 8.5×10^{-3} L

(5) 776,000 g (a) 776,000 is the numerical value, g is unit (b) 3 significant figures (c) 77 known, 6 estimated (dd) 780,000 g to two significant figures (e) 7.8×10^5 g

CHAPTER THREE

Classification of Matter

SELF-EVALUATION SECTION

1. All matter in the universe can be classified into one of three states — gas, liquid, or solid. Determine to which of the three states of matter each of the following descriptions relates.

	Description	States of Matter
(1)	Has a definite fixed shape.	_____
(2)	Particles flow over each other while retaining fixed volume.	_____
(3)	Exerts a pressure on all walls of the container.	_____
(4)	Exhibits no or very slight compressibility.	_____
(5)	Particles move independently of each other.	_____
(6)	Particles arranged in regular, fixed geometric pattern.	_____
(7)	Exhibits slight compressibility.	_____
(8)	Exhibits very high compressibility.	_____

2. Examine the following list of items. Some are mixtures, others are elements or compounds. Place an *S* next to the pure substance and an *M* next to the mixtures.

(1)	wood	_____	(6)	water	_____
(2)	sodium chloride	_____	(7)	air	_____
(3)	milk	_____	(8)	chlorine	_____
(4)	oxygen	_____	(9)	soil	_____
(5)	rubber	_____	(10)	gasoline	_____

3. Fill in with a response appropriate to Chapter 3.

There are 111 presently known (1) _____, of which at least 88 occur naturally on earth. These building-block units can be combined by means of chemical reactions to form a multitude of chemical (2) _____. Thus, one can see that compounds composed of two or more (3) _____ can be formed or broken down into the constituent (4) _____ by various reactions. When an element is subdivided into smaller and smaller pieces, the final indivisible particle that is left is called an (5) _____. Atoms can combine in two ways in simple whole number ratios. Atoms form small, uncharged units called (6) _____. An example is sugar, which can be subdivided into smaller and smaller particles until we finally obtain one (7) _____ of sugar. We cannot divide this unit further without destroying its identity. In other compounds, atoms exist as electrically charges species called (8) _____. These charged species may be composed of one or several atoms. If the ion is positively charged (+), it is called a (9) _____. It the ion is negatively charged (−), it is called a (10) _____. Compounds consisting of ions do not form (11) _____. The formula for an ionic compound represents the smallest whole number ratio of ions present in the crystalline structure.

4. Write the chemical symbol for the five most abundant elements in the earth's crust, seawater, and atmosphere. Also write the six most abundant elements in the human body. Which elements are common to both lists?

Earth's Chemicals	Human Body
a. _____	a. _____
b. _____	b. _____
c. _____	c. _____
d. _____	d. _____
e. _____	e. _____
	f. _____

Elements common to both:

5. Write the correct chemical symbols for the elements listed in Column A. Likewise, write out the correct chemical names for the symbols listed in Column B. Study Table 3.5 in the text thoroughly before writing this exercise.

Column A		Column B	
Aluminum	_____	Ar	_____
Antimony	_____	Ba	_____
Arsenic	_____	Bi	_____
Boron	_____	Br	_____
Cadmium	_____	Cl	_____
Calcium	_____	Cr	_____
Carbon	_____	Co	_____
Copper	_____	F	_____
Gold	_____	I	_____
Helium	_____	Mn	_____
Iron	_____	Ne	_____
Lead	_____	Ni	_____
Magnesium	_____	P	_____
Mercury	_____	Si	_____
Nitrogen	_____	Na	_____
Potassium	_____	S	_____
Silver	_____	Ti	_____
Tin	_____	U	_____
Tungsten	_____		
Zinc	_____		

6. Look at the following list of chemicals. From what you've learned in Chapter 3, try to identify each formula as to whether it represents a compound or not, and if the formula is that of a compound whether the compound is molecular or ionic in nature as described in Chapter 3.

Formula	Compound? Yes or No	If yes then	Molecular or Ionic
(1) I_2	_____		_____ _____
(2) CO	_____		_____ _____
(3) P_4	_____		_____ _____
(4) NaCl	_____		_____ _____
(5) H_2O	_____		_____ _____

Formula	Compound? Yes or No	If yes then Molecular or Ionic	
(6) CH$_4$	_____	_____	_____
(7) Fe	_____	_____	_____
(8) Co	_____	_____	_____
(9) He	_____	_____	_____
(10) NH$_3$	_____	_____	_____

7. Circle the appropriate response.

It is often difficult to tell if a material we have obtained in an experiment is a pure substance, such as a compound, or a mixture of compounds. If the characteristics that distinguish a compound from a mixture are understood, sometimes a decision can be made or an experiment performed to tell if we, in fact, have a pure compound. A mixture may be composed of compounds or elements in (1) varying/fixed proportions. The component elements of a pure chemical compound can be obtained only by (2) physical/chemical changes while a mixture can be separated by (3) physical/chemical means. Identification of the components of a mixture shows that each individual fraction (4) has/has not lost its identity upon forming a mixture while a chemical compound (5) does/ does not resemble the component elements.

Indicate by a *C* or an *M* whether the following statements refer to a *compound* or a *mixture*.

(6) Composed of elements, compounds, or both in variable composition _____

(7) Components do not lose their identity _____

(8) Components separated by chemical changes only _____

(9) Does not resemble elements from which it was formed _____

(10) Composed of two or more elements in fixed proportion by mass _____

(11) Separation may often be made by simple physical or mechanical means _____

8. Fill in the blank space or circle the appropriate response.

The 111 elements, both naturally occurring and artificial, can be classified into three subgroups: metals, nonmetals, and metalloids. Metals, except for mecury, are in the (1) _____ state at room temperature. They have a (2) dull/lustrous appearance and are (3) good/poor conductors of (4) _____ and (5) _____ . Metals are also (6) _____ which means they can be drawn into wires, and are (7) _____ , which means they can be flattened into sheets. Most metals have (8) _____ melting points and (9) _____ densities. Nonmetals, in general, have properties (10) similar/opposite to those of metals while elements such as silicon and arsenic, which are called (11) _____, have properties resembling both metals and nonmetals. Metallic elements tend to react with (12) _____ elements to form compounds while non-metallic elements will react with any of the three subgroups. Seven of the non-metallic elements occur naturally as diatomic molecules. Diatomic means that (13) _____ atoms are joined together in chemical bond. Most of the diatomic elements are (14) _____, but one is a liquid and one is a solid at room temperature. Write the formulas for as many of the diatomic elements as you can (15) _____ .

9. How many atoms of nitrogen are contained in each of the following formulas?
   (1) $NH_4S_2O_8$ _____
   (2) NH_4CNS _____
   (3) $Na_3Co(NO_2)_6$ _____
   (4) $Na_2Fe(CN)_5NO \cdot 2 H_2O$ _____
   (5) $Fe_2(SO_4)_3(NH_4)_2SO_4 \cdot 24 H_2O$ _____
   (6) $K_3Fe(CN)_6$ _____
   (7) $NH_4C_2H_3O_2$ _____

RECAP SECTION

Chapter 3 has given you an introduction to the vocabulary of the chemist. We have classified matter into substances and mixtures, then distinguished between elements and compounds. The new terms of atom, molecule, cation, and anion were also added to our vocabulary. We have learned to identify the physical states of matter, to distinguish among metals, nonmetals, and metalloids, and to understand the meaning of a chemical formula. With this new understanding of the language of chemistry you are ready to study the properties of matter.

ANSWERS TO QUESTIONS AND SOLUTIONS TO PROBLEMS

1. (1) solid (2) liquid (3) gas (4) solid
   (5) gas (6) solid (7) liquid (8) gas

2. (1) M (2) S (3) M (4) S
   (5) M (6) S (7) M (8) S
   (9) M (10) M

3. (1) elements (2) compounds (3) elements (4) elements
   (5) atom (6) molecules (7) molecule (8) ions
   (9) cation (10) anion (11) molecules

4. Earth: O, Si, Al, Fe, Ca
   Body: O, C, H, N, Ca, P
   Oxygen and calcium are common to both.

5. Aluminum – Al   Ar – Argon
   Antimony – Sb   Ba – Barium
   Arsenic – As   Bi – Bismuth
   Boron – B   Br – Bromine
   Cadmium – Cd   Cl – Chlorine
   Carbon – C   Cr – Chromium
   Copper – Cu   F – Fluorine
   Gold – Au   I – Iodine
   Helium – He   Mn – Manganese
   Iron – Fe   Ne – Neon
   Lead – Pb   Ni – Nickel
   Magnesium – Mg   P – Phosphorus
   Mercury – Hg   Si – Silicon
   Nitrogen – N   Na – Sodium
   Potassium – K   S – Sulfur
   Silver – Ag   Ti – Titanium
   Tin – Sn   U – Uranium
   Tungsten – W
   Zinc – Zn

6. (1) No (2) Yes, molecular (3) No (4) Yes, ionic (5) Yes, molecular (6) Yes, molecular (7) No (8) No (9) No (10) Yes, molecular

7. (1) Varying (2) chemical
   (3) physical (4) has not
   (5) does not (6) M
   (7) M (8) C
   (9) C (10) C
   (11) M

8.
   - (1) solid
   - (2) lustrous
   - (3) good
   - (4) heat
   - (5) electricity
   - (6) ductile
   - (7) malleable
   - (8) high
   - (9) high
   - (10) opposite
   - (11) metalloids
   - (12) nonmetallic
   - (13) two
   - (14) gases
   - (15) $H_2, N_2, O_2, F_2, Cl_2, Br_2, I_2$

9. (1) 1   (2) 2   (3) 6   (4) 6   (5) 2   (6) 6   (7) 1

CHAPTER FOUR

Properties of Matter

SELF-EVALUATION SECTION

1. Statements about the properties and changes of various substances are given below. Place the appropriate letter of identification in each space provided.

   a. physical property   b. chemical property
   c. physical change   d. chemical change

Description	Identification
(1) Potassium is solid at room temperature.	
(2) Potassium is among the most reactive elements in nature.	
(3) The melting point of potassium is 63.6°C.	
(4) Potassium reacts vigorously with water.	
(5) Potassium added to water produces hydrogen gas plus a water solution of potassium hydroxide.	
(6) At 760°C, potassium begins to boil and to change from a liquid to a gas.	
(7) At high temperatures, potassium reacts with hydrogen gas to form a metallic hydride.	
(8) Alkali metals, such as potassium, form salts with many other elements.	
(9) Potassium metal is shiny and fairly soft.	
(10) Potassium is useful in the manufacture of fertilizer compounds.	

2. Distinguish between *potential* and *kinetic* energy.

After this section, we will be able to define the two types of energy that matter possesses and to distinguish between them. Energy is simply defined as the capacity to do work.

Potential energy is the energy which matter has as a result of chemical bonds or because of its position in relationship to another place; kinetic energy is strictly the energy of motion. Filling in the items in this story about a bike rider may help to make the difference between potential energy and kinetic energy clearer to you. Use the symbols *P* (for *potential*) and *K* (for *kinetic*) in the spaces provided to identify the type of energy described by the underlined word or within the statement.

One fine summer morning, as the birds were chirping and flitting (1) _____ about his window, Michael awoke and thought about a bike ride. He scrambled (2) _____ out of bed, dressed, and hurried to breakfast. His breakfast of fruit, eggs, toast, and milk was a source of (3) _____ energy for his ride. He wheeled (4) _____ his bike out of the garage and began converting the (5) _____ energy of his breakfast into (6) _____ energy as he pumped up the hill. As he pedaled, his foot had greater (7) _____ energy at the top of the sprocket than it had at the bottom. As his foot started down, this higher (8) _____ energy was converted into (9) _____ energy. As Michael picked up speed, his bike increased in (10) _____ energy. After pedaling vigorously, he reached the top of the hill and paused. In relationship to the bottom of the hill, he and his bike had increased in (11) _____ energy. As he turned around and started down the hill, he began to convert this (12) _____ energy into (13) _____ energy. All Michael had to do was coast back down and enjoy the ride.

3. Determine whether the following statements or blanks relate to *chemical, light, heat, mechanical,* or *electrical* energy.

Radiant energy <u>(1)</u> from the sun is used in the (1) _____
process of photosynthesis.
Photosynthesis involves a conversion of radiant
energy into the <u>(2)</u> energy of sugars, starch,
and other compounds.   (2) _____
Organisms utilize the <u>(3)</u> energy of foodstuffs for movement and activity   (3) _____
and, in the case of warm-blooded animals, for
maintaining body temperature <u>(4)</u> .   (4) _____
When wood is burned, the two forms of energy   (5) _____
that are released are <u>(5)</u> and <u>(6)</u> .   (6) _____
The chemical energy of a car battery is con-
verted into <u>(7)</u> energy, which turns the   (7) _____
starter motor.
Fuels such as gasoline contain <u>(8)</u> energy   (8) _____
which, through the process of combustion, is
converted to the <u>(9)</u> energy of moving pis-   (9) _____
tons, valves, gears, and wheels.

Hydroelectric dams use the energy of falling water to turn turbines that convert __(10)__ energy into __(11)__ energy.

(10) _____

(11) _____

4. As our last section indicates, the various forms of energy can be converted from one form to another; nothing was said about gaining energy or losing energy during these changes. Various careful experiments have shown that energy cannot be created or destroyed; it can only be transformed from one form to another. The constancy of energy is summarized as the Law of Conservation of Energy. A little earlier in the text, you read that during ordinary chemical reactions the total mass of the reactants was equal to the total mass of the products. Thus, during chemical reactions, mass is neither created nor destroyed. The summarizing statement concerning this experimental fact is called the Law of Conservation of Mass.

5. When we talk about the amount of heat energy involved in a chemical process, we use the SI unit of measurement called the *joule*. We define the joule by an experimental procedure – a common practice in science. If we measure out 1 gram of water and place it at 14.5° Celsius, then 4.184 joules of heat energy from a match, gas burner, or other source will raise the temperature of the 1 gram of water 1° Celsius to 15.5° Celsius. In words, the number of joules gained or lost is equal to the mass of water times the temperature change times a constant. The constant is the *specific heat* for the material in question. For water, the specific heat is 4.184 J/g°C. You will work some sample problems later.

Remembering that the number of joules (abbreviated J) gained or lost during heat energy changes is equal to the mass of water times the specific heat of the substance times the temperature difference, we can write the following equation.

$$\text{joules} = (\text{grams of water})(\text{specific heat of substance})(\Delta t)$$

The symbol Δt means "temperature difference". The last term in the equation is the specific heat for the substance in question. For water, the specific heat is 4.184 J/g°C. An equation with units attached would be

$$\text{joules} = g \times 4.184 \text{ J/g°C} \times °C$$

Try the following problems:

(1) How many joules of heat energy are required to heat 330 g of water from 25°C to boiling (100°C)?

Do your calculations here.

(2) How many grams of water can be heated from 5°C to 52°C with 1.0×10^2 joules of heat energy?

Do your calculations here.

(3) Let's work problems (1) and (2) again but this time we will use the calorie unit of heat measurement. Recall that the specific heat of water in calorie units is 1 cal/g°C or in words "one calorie of heat energy will raise one gram of water one degree Celsius". Therefore, how many calories of heat energy are required to heat 330 g of water from 25°C to boiling (100°C)?

Do your calculations here.

(4) In a similar manner, how many grams of water can be heated from 5°C to 52°C with 1.0×10^2 cal of heat energy?

Do your calculations here.

Challenge Problems

6. 20.0 g of H_2 and 20.0 of O_2 are reacted to produce H_2O by the following equation:

    $2 H_2 + O_2 \rightarrow 2 H_2O$

    After the reaction, 17.5 g of H_2 are left. What is the precent composition of H_2 and O_2 in H_2O?

    Do your calculations here.

7. Concentrated sulfuric acid (H_2SO_4) has a density of 1.84 g/mL. Dilute sulfuric acid has a density of 1.18 g/mL. For a laboratory experiment you need a sulfuric acid solution with a density of 1.50 g.mL. Using 50.0 mL of 1.84 g/mL sulfuric acid, how many mL of 1.18 g/mL acid will you need to mix in to obtain the final mixture of 1.50 g/mL?

Do your calculations here.

8. A 50. g piece of copper at 212°F is dropped into 100. g of ethyl alcohol at 72°F. The specific heat of copper is 0.0921 $\frac{cal}{g°C}$, that of ethyl alcohol is 0.511 $\frac{cal}{g°C}$. Calculate the final temperature of the mixture.

Do your calculations here.

RECAP SECTION

In Chapter 4 we examined the physical and chemical properties of matter. Physical properties can be determined without altering the substance, while chemical properties involve the interaction of substances resulting in different material. Whether matter undergoes a physical change or a chemical change, energy is conserved. Energy can take many forms, including the heat and light most commonly seen during chemical changes. The material in these first chapters has given you the foundation to begin to explore the atom and its internal structure.

ANSWERS TO QUESTIONS AND SOLUTIONS TO PROBLEMS

1. (1) a   (2) b   (3) a   (4) b
   (5) d   (6) c   (7) d   (8) b
   (9) a   (10) b

2. (1) K   (2) K   (3) P   (4) K
   (5) P   (6) K   (7) P   (8) P
   (9) K   (10) K   (11) P   (12) P
   (13) K

3. (1) light   (2) chemical   (3) mechanical   (4) heat
   (5) heat   (6) light   (7) electrical   (8) chemical
   (9) mechanical   (10) mechanical   (11) electrical

4. No answers necessary.

5. (1) 100,000 joules or 1.0×10^5 joules to two significant figures. We know the mass of water (330 g) and can determine the temperature difference by subtraction. The unknown quantity is the number of joules of heat energy required. Make the necessary substitutions in the equation.

   $$\text{joules} = 330 \text{ g} \times 4.184 \text{ J/g°C} \times (100°C - 25°C)$$
   $$= 330 \text{ g} \times 4.184 \text{ J/g°C} \times 75°C$$
   $$= 103{,}554 \text{ joules}$$
   $$= 1.0 \times 10^5 \text{ joules to 2 significant figures}$$

   (2) 5.1 g of water (2 significant figures)
   We are asked to determine the mass of water which 1.0×10^2 joules of heat energy can raise from 5°C to 52°C.
   Solving our equation for grams we have

   $$\text{joules} = m \times \text{specific heat} \times \Delta t$$
   $$m = \frac{\text{joules}}{\text{specific heat} \times \Delta t}$$
   $$= \frac{1.0 \times 10^2 \text{ joules}}{4.184 \text{ J/g°C} \times (52°C - 5°C)}$$
   $$= \frac{1000 \text{ joules}}{4.184 \text{ J/g°C} \times 47°C}$$
   $$= 5.1 \text{ g of water}$$

   (3) 25,000 cal to 2 significant figures
   We know the mass of water (330 g) and know the temperature difference by subtraction. The unknown quantity again is the number of calories of heat energy required. Make the necessary substitutions in the equation.

$$\text{cal} = 330 \text{ g} \times 1 \text{ cal/g°C} \times (100°C - 25°C)$$
$$= 330 \not{g} \times 1 \text{ cal}/\not{g}\not{°C} \times 75\not{°C}$$
$$= 24{,}750 \text{ cal}$$
$$= 2.5 \times 10^4 \text{ cal (2 significant figures)}$$

(4) 21 g of water

Again, what mass of water can 1.0×10^2 cal raise from 5°C to 52°C?

Solving our equation for grams we will have

$$\text{cal} = m \times \text{specific heat} \ \Delta t$$
$$m = \frac{\text{cal}}{\text{specific heat} \times \Delta t}$$
$$= \frac{1.0 \times 10^2 \text{ cal}}{1 \text{ cal/g°C} \times (52°C - 5°C)}$$
$$= \frac{1.0 \times 10^2 \ \not{\text{cal}}}{1 \ \not{\text{cal}}/g\not{°C} \times 47\not{°C}}$$
$$= 21 \text{ g of water}$$

6. 11.1% H_2 and 88.9% O_2

From the data given, the mass of water formed is the amount of O_2 present plus the amount of H_2 reacted.

$$20.0 \text{ g } O_2 + 2.5 \text{ g } H_2 = 22.5 \text{ g } H_2O$$

$$\% \ H_2 = \frac{\text{mass } H_2}{\text{mass } H_2O} \times 100 = \frac{2.5 \not{g}}{22.5 \not{g}} \times 100 = 11.1\%$$

$$\% \ O_2 = \frac{\text{mass } O_2}{\text{mass } H_2O} \times 100 = \frac{20.0 \not{g}}{22.5 \not{g}} \times 100 = 88.9\%$$

7. 53 mL of 1.18 g/mL sulfuric acid required. Using the density equals mass divided by volume equation we can solve the problem either by looking at the masses of sulfuric acid involved or looking at the densities. Let's use the mass approach.

We can say that the following relationship exists:

(g of concentrated H_2SO_4) + (g of dilute H_2SO_4) = g of mixture and

$$\text{density} = \frac{\text{mass}}{\text{volume}} \text{ or mass} = (\text{density})(\text{volume})$$

Let x = mL of 1.18 g/mL of sulfuric acid.

Therefore g of con. $H_2SO_4 = (1.84 \text{ g}/\not{\text{mL}})(50. \ \not{\text{mL}}) = 92 \text{ g}$
g of dil. $H_2SO_4 = (1.18 \text{ g}/\not{\text{mL}})(x \ \not{\text{mL}}) = 1.18 \ x \text{ g}$

This term contains the unknown, i.e., how much dilute H_2SO_4 is needed.

The density of the mixture is to be 1.50 g/mL so we can set up the following equation:

$$(1.84 \text{ g/mL} \times 50. \text{ mL}) + (1.18 \text{ g/mL} \times x) = \underbrace{(50 \text{ mL} + x)}_{\text{(final volume)}} \times \underbrace{1.50 \text{ g/mL}}_{\text{(final density)}}$$

Multiply, collect like terms

$$92 \text{ mL} + 1.18(x) = 75 \text{ mL} + 1.50(x)$$
$$17 \text{ mL} = 0.32(x)$$
$$x = \frac{17}{0.32} \text{ mL} = 53 \text{ mL}$$

8. In examining the problem we should notice first that the temperatures are given in °F while specific heats contain the temperature unit of °C. In general, as we work with chemical problems we will find that conversions are necessary when we see temperatures given as °F. Therefore, the copper metal's temperature is

$$°C = \frac{(°F - 32)}{1.8} = \frac{(212 - 32)}{1.8} = 100.°C$$

The temperature of the ethyl alcohol is

$$°C = \frac{(°F - 32)}{1.8} = \frac{(72 - 32)}{1.8} = 22°C$$

Next we can say that the copper metal piece has been heated up, dropped into a cooler material, the alcohol, and caused the final mixture to increase in temperature. The calories of heat gained by the copper as it was heated are lost to the alcohol so that the final temperatures of the copper and alcohol are equal.

Another way to state this is that the heat lost by the copper is equal to the heat gained by the alcohol.

Mathematically, the equality would be:

heat lost$_{Cu}$ = heat gained$_{alcohol}$

heat lost$_{Cu}$ = $m_{Cu} C_{Cu} \Delta T_{Cu}$ = $(50. \text{ g})\left(0.0921 \frac{\text{cal}}{\text{g°C}}\right)(100.°C - T_f)$

Heat gained$_{alcohol}$ = $m_{alcohol} C_{alcohol} \Delta T_{alcohol}$ = $(100 \text{ g})\left(0.511 \frac{\text{cal}}{\text{g°C}}\right)(T_f - 22°C)$

Notice that T_f is second in ΔT_{Cu} and first in $\Delta T_{alcohol}$. This is a result of the Cu being cooled (to a lower temperature) while the alcohol is being warmed (to a higher temperature).

Since heat lost$_{Cu}$ = heat gained$_{alcohol}$

$$(50. \text{ g})\left(0.0921 \frac{\text{cal}}{\text{g°C}}\right)(100.°C - T_f) = (100.\text{g})\left(0.511 \frac{\text{cal}}{\text{g°C}}\right)(T_f - 22°C)$$

Upon expansion,

$$460.5 \text{ cal} - \left(4.605 \frac{\text{cal}}{°C}\right) T_f = \left(51.1 \frac{\text{cal}}{°C}\right) T_f - 1124.2 \text{ cal}$$
$$1584.7 \text{ cal} = \left(55.705 \frac{\text{cal}}{°C}\right) T_f$$
$$28°C = T_f$$

CHAPTER FIVE

Early Atomic Theory and Structure

SELF-EVALUATION SECTION

1. Fill in the blank space or circle the appropriate response.

   When a chemist analyzes a chemical compound, the percentage composition of the compound by mass is found to be constant. For example, water always contains 11.2% hydrogen and 88.8% oxygen. This experimental observation illustrates the Law of (1) _____ _____. The Law of Definite Composition states that the percentage composition of a pure compound is (2) _____.

   As we begin to characterize various elements, one definition should automatically come to mind. The number of protons in an atom of any element is called the (3) _____ _____. This number is also the same as the number of (4) _____ in a neutral atom. This is logical since in an electrically neutral atom the number of positive charges has to equal the number of (5) _____ charges.

   The two subatomic particles which have a mass of 1 amu are the (6) _____ and (7) _____. The negatively charge particle, called an (8) _____, has an assigned mass of (9) _____. Thus, essentially all of the mass of an atom is located in the (10) _____ of an atom. In describing the relative masses of the atoms of various elements, we use the term (11) _____, which represents the mass of the number of protons and neutrons in an atom. However, as we look at a list of relative atomic masses for various elements, we find that most values are not whole numbers. This may seem odd since the system of atomic masses is based on a particular isotope of the element, (12) _____, having an exact atomic mass of 12.00. We must remember that atoms of an element can have different numbers of neutrons in the nucleus and still be the same element. Thus, a carbon atom can have six neutrons or eight neutrons in the nucleus. The atom is still carbon since the atomic number remains 6, but the relative mass of the two atoms is different. And, depending on the relative abundance of the various atoms with differing numbers of neutrons, the average atomic mass of a sample of the element will not be a whole number. The atomic mass of the element is a weighted average of all the atoms present in the sample.

Atoms of a particular element which have the same atomic (13) _____ but differ in atomic mass are called (14) _____ of that element. Some radioactive isotopes are extremely important today. Some such as $^{131}_{53}I$, $^{60}_{27}Cu$, or $^{32}_{15}P$ are used in medical applications. Others, such as $^{90}_{38}Sr$, are environmental contaminants and may pose a long-term problem for various organisms. The simplest element, hydrogen, has three isotopes. They are named (15) _____, (16) _____, and (17) _____. The symbols for the elements used in the preceding sentences have two numbers in addition to the chemical symbol. The upper number, which is the larger, is the (18) _____ _____. The lower number is the (19) _____ _____.

2. Indicate by a yes or no response which of the following statements are generally accepted today.

   (1) Atoms of the same element may vary in mass but are the same size. _____
   (2) Compounds are formed by the union or joining of two or more atoms of different element. _____
   (3) Atoms of different elements have different masses and sizes. _____
   (4) Atoms of two elements may combine in different ratios to form more than one compound. _____
   (5) Elements are composed of small, indivisible units called electrons. _____
   (6) In forming compounds, atoms can combine in fractional units. _____
   (7) The fundamental building blocks of elements are small particles called atoms. _____
   (8) Atoms of the same element are alike in mass and size. _____
   (9) In forming compounds, atoms combine in small whole-number ratios such as one to one, one to two, and so forth. _____
   (10) Atoms of two elements can combine in only one fixed ratio to form only one compound. _____

3. Description of properties of subatomic particles. Fill in the spaces of the table with the correct response from the list of possible responses below.

Particle	Symbol	Relative Mass (amu)	Charge	Location in Atom
Electron				
Proton				
Neutron				

Possible responses

Symbol	Relative Mass (amu)	Charge	Location in Atom
	+2	−1	nucleus
p	+1	−2	outside nucleus
	−1	+1	
e	$-\frac{1}{2}$	−2	
n	0	0	
	$\frac{1}{1837}$	$-\frac{1}{2}$	

4. Fill in the blank space or circle the appropriate response.

The three principal subatomic particles which are of interest to chemists are the (1) _____, (2) _____, and (3) _____. A series of experiments after the turn of the century showed that protons and (4) _____ were found in the (5) _____ of an atom while the (6) _____ were in regions outside the nucleus. Rutherford demonstrated that practically all of the mass of an atom is to be found in the (7) _____ and that atoms are mostly empty space. The experiments of Rutherford and others left unanswered questions about the nature of electrons in atoms.

5. Atomic masses, atomic numbers, and number of neutrons.

   (1) Determine the number of neutrons in an atom of each of the following elements.

   $^{23}_{11}Na$ _____ $^{65}_{30}Zn$ _____

   $^{115}_{48}Cd$ _____ $^{40}_{18}Ar$ _____

   $^{35}_{17}Cl$ _____ $^{40}_{20}Ca$ _____

(2) Determine the approximate *atomic mass* for an atom of each of the following isotopes.

	Atomic Number	Number of Neutrons	Atomic Mass
Li	3	4	_____
K	19	20	_____
Hg	80	121	_____
Fe	26	30	_____
H	1	0	_____
P	15	16	_____

6. Which of the following statements are true about the neutral atoms $^{12}_{6}C$ and $^{14}_{6}C$?

(1) $^{12}_{6}C$ and $^{14}_{6}C$ are isotopes.

(2) $^{12}_{6}C$ has two less electrons than $^{14}_{6}C$.

(3) $^{14}_{6}C$ has two more protons than $^{12}_{6}C$.

(4) $^{12}_{6}C$ has two less neutrons than $^{14}_{6}C$.

RECAP SECTION

Chapter 5 provides our first look inside the atom. Early models of the atom were modified to incorporate the concept of electric charge. The atom contains three types of subatomic particles, the proton, the neutron, and the electron. The arrangement of these particles within the atom was proposed by Ernest Rutherford. In his model of the atom the neutrons and the protons are located within the nucleus, while the electrons occupy the remainder of the atom. This brief glimpse into atomic structure allows us to relate the atomic number to the number of protons in the nucleus (they are equal). These early models of the atom also led to the discovery that isotopes are atoms of the same element with differing numbers of neutrons. From here we can begin to look at the composition of compounds.

ANSWERS TO QUESTIONS AND SOLUTIONS TO PROBLEMS

1.
   - (1) Definite Composition
   - (2) constant
   - (3) atomic number
   - (4) electrons
   - (5) negative
   - (6) proton
   - (7) neutron
   - (8) electron
   - (9) 1/1837 amu
   - (10) nucleus
   - (11) atomic mass
   - (12) carbon
   - (13) number
   - (14) isotopes
   - (15) protium
   - (16) deuterium
   - (17) tritium
   - (18) mass number
   - (19) atomic number

2.
   - (1) yes
   - (2) yes
   - (3) yes
   - (4) yes
   - (5) no
   - (6) no
   - (7) yes
   - (8) no
   - (9) yes
   - (10) no

3.

Particle	Symbol	Relative Mass (amu)	Charge	Location in Atom
Electron	e	1/1837	−1	outside nucleus
Proton	p	1	+1	nucleus
Neutron	n	1	0	nucleus

4.
   - (1) protons
   - (2) neutrons
   - (3) electrons
   - (4) neutrons
   - (5) nucleus
   - (6) electrons
   - (7) nucleus
   - (8) higher

5.
   - (1) Na = 12, Zn = 35, Cd = 67, Ar = 22, Cl = 18, Ca = 20
   - (2) Li = 7, Fe = 56, K = 39, H = 1, Hg = 201, P = 31

6.
   - (1) T
   - (2) F
   - (3) F
   - (4) T

CHAPTER SIX

Nomenclature of Inorganic Compounds

SELF-EVALUATION SECTION

1. Write the correct names for the chemical formulas as listed. For additional assistance in naming compounds, refer to Tables 6.2, 6.3, 6.4, and 6.6.

    (1)　KF　　　　　K_2O　　　　　KNO_3　　　　　K_3PO_4

    _____

    (2)　MgF_2　　　MgO　　　　　$Mg(NO_3)_2$　　　$Mg_3(PO_4)_2$

    _____

    (3)　FeF_3　　　Fe_2O_3　　　$Fe(NO_3)_3$　　　$FePO_4$

    _____

    (4)　HF　　　　　H_2O　　　　　HNO_3　　　　　H_3PO_4

    _____

    (5)　SnF_4　　　SnO_2　　　　$Sn(NO_3)_4$　　　$Sn_3(PO_4)_4$

    _____

2. Write the names of the following compounds, which are composed of non-metallic elements covalently bonded together.

    (1)　CO　　　　　CO_2　　　　　PCl_5　　　　　P_2O_5

    _____

    (2)　CCl_4　　　NO　　　　　　S_2Cl_2　　　　　N_2O_3

    _____

– 41 –

3. Write the correct names for these common compounds.

   (1) Common acids

   HCl   H$_2$SO$_4$ HC$_2$H$_3$O$_2$ H$_2$CO$_3$
   _____   _____   _____   _____

   (2) Common salts

   MgSO$_4$ K$_2$CO$_3$ NaNO$_3$ Ca$_3$(PO$_4$)$_2$
   _____   _____   _____   _____

   (3) Common oxides

   Al$_2$O$_3$ PbO   N$_2$O   CaO
   _____   _____   _____   _____

   (4) Common chlorides

   NH$_4$Cl   NaCl   KCl   MgCl$_2$
   _____   _____   _____   _____

4. Write the names for the formulas and the formulas for the written names listed below. Refer to the tables in Chapter 6.

   (1) AgCl   _____
   (2) KNO$_3$ _____
   (3) SnCl$_4$ _____
   (4) Al$_2$O$_3$ _____
   (5) SO$_3$ _____
   (6) NH$_4$NO$_3$ _____
   (7) AlPO$_4$ _____
   (8) Na$_2$SO$_4$ _____
   (9) (NH$_4$)$_2$CO$_3$ _____
   (10) NaI   _____
   (11) FeCl$_2$ _____
   (12) BaO   _____
   (13) K$_2$Cr$_2$O   _____
   (14) NaHCO$_3$ _____
   (15) CO   _____
   (16) Hydrochloric acid   _____
   (17) Lead (II) nitrate   _____
   (18) Potassium permanganate _____

(19) Magnesium chloride _____

(20) Carbon dioxide _____

(21) Barium sulfate _____

(22) Iron (III) Chloride _____

(23) Silicon dioxide _____

(24) Sulfuric acid _____

(25) Dinitrogen tetroxide _____

(26) Nitric acid _____

(27) Cobalt (II) chloride _____

(28) Sodium sulfite _____

(29) Hydrosulfuric acid _____

(30) Carbon tetrachloride _____

5. Give the formulas for the following salts, which contain more than one positive ion.

  (1) Sodium hydrogen carbonate

  (2) Magnesium ammonium phosphate

  (3) Sodium potassium sulfate

  (4) Ammonium hydrogen sulfide

  (5) Potassium hydrogen sulfate

  (6) Potassium aluminum sulfate

7. Write the correct formulas for compounds formed by matching each cation with all anions.

	F^-	O^{2-}	NO_3^-	PO_4^{3-}
K^+	KF	K_2O	KNO_3	K_3PO_4
Mg^{2+}	MgF_2	MgO	$Mg(NO_3)_2$	$Mg_3(PO_4)_2$
Fe^{3+}	FeF_3	Fe_2O_3	$Fe(NO_3)_3$	$FePO_4$
H^+	HF	H_2O	HNO_3	H_3PO_4
Cu^+	CuF	Cu_2O	$CuNO_3$	Cu_3PO_4
Sn^{4+}	SnF_4	SnO_2	$Sn(NO_3)_4$	$Sn_3(PO_4)_4$
Ca^{2+}	CaF_2	CaO	$Ca(NO_3)_2$	$Ca_3(PO_4)_2$
Al^{3+}	AlF_3	Al_2O_3	$Al(NO_3)_3$	$AlPO_4$

RECAP SECTION

Chapter 6 is one of the self-contained study units in the text for you to refer to as needed. After completing the material in the text, you will be familiar with naming binary compounds composed of a metal and a nonmetal and those composed of two nonmetals. It is common to find oxygen-containing radicals in polyatomic compounds. Rules are given for naming these compounds containing polyatomic ions. Only through practice and experience in a laboratory will you become adept at using chemical names. Try to use names whenever possible. Ask your instructor the name of any chemical you are not sure about. It is a good practice to have the original chemical bottles present during a lab period even though your instructor may provide the chemicals in solution. Ask to see the reagent bottles and read labels carefully.

ANSWERS TO QUESTIONS

1. (1) Potassium fluoride
Potassium oxide
Potassium nitrate
Potassium phosphate

    (2) Magnesium fluoride
Magnesium oxide
Magnesium nitrate
Magnesium phosphate

    (3) Iron (III) fluoride
Iron (III) oxide
Iron (III) nitrate
Iron (III) phosphate

    (4) Hydrogen fluoride
Water
Nitric acid
Phosphoric acid

    (5) Tin (IV) fluoride
Tin (IV) oxide
Tin (IV) nitrate
Tin (IV) phosphate

2. (1) Carbon monoxide
Carbon dioxide
Phosphorus pentachloride
Diphosphorus pentoxide

    (2) Carbon tetrachloride
Nitrogen oxide
Disulfur dichloride
Dinitrogen trioxide

3. (1) Hydrochloric acid
Sulfuric acid
Acetic acid
Carbonic acid

    (2) Magnesium sulfate
Potassium carbonate
Sodium nitrate
Calcium phosphate

    (3) Aluminum oxide
Lead (II) oxide
Dinitrogen oxide
Calcium oxide

    (4) Ammonium chloride
Sodium chloride
Potassium chloride
Magnesium chloride

4. (1) Silver chloride
   (2) Potassium nitrate
   (3) Tin (IV) chloride
   (4) Aluminum oxide
   (5) Silver trioxide
   (6) Ammonium nitrate
   (7) Aluminum phosphate
   (8) Sodium sulfate
   (9) Ammonium carbonate
   (10) Sodium iodide
   (11) Iron (III) chloride
   (12) Barium oxide
   (13) Potassium dichromate
   (14) Sodium hydrogen carbonate
   (15) Carbon monoxide
   (16) HCl
   (17) $Pb(NO_3)_2$
   (18) $KMnO_4$
   (19) $MgCl_2$
   (20) CO_2
   (21) $BaSO_4$
   (22) $FeCl_3$
   (23) SiO_2
   (24) H_2SO_4
   (25) N_2O_4
   (26) HNO_3
   (27) $CoCl_2$
   (28) Na_2SO_3
   (29) H_2S
   (30) CCl_4

5. (1) $NaHCO_3$
   (2) $MgNH_4PO_4$
   (3) $NaKSO_4$
   (4) NH_4HS
   (5) $KHSO_4$
   (6) $KAl(SO_4)_2$

6.
KF	K$_2$O	KNO$_3$	K$_3$PO$_4$
MgF$_2$	MgO	Mg(NO$_3$)$_2$	Mg$_3$(PO$_4$)$_2$
FeF$_3$	Fe$_2$O$_3$	Fe(NO$_3$)$_3$	FePO$_4$
HF	H$_2$O	HNO$_3$	H$_3$PO$_4$
CuF	Cu$_2$O	CuNO$_3$	Cu$_3$Po$_4$
SnF$_4$	SnO$_2$	Sn(NO$_3$)$_4$	Sn$_3$(PO$_4$)$_4$
CaF$_2$	CaO	Ca(NO$_3$)$_2$	Ca$_3$(PO$_4$)$_2$
AlF$_3$	Al$_2$O$_3$	Al(NO$_3$)$_3$	AlPO$_4$

CHAPTER SEVEN

Quantitative Composition of Compounds

SELF-EVALUATION SECTION

1. In a chemical laboratory, masses are usually determined on a balance in grams or milligrams. It would be impossible to find the mass of one atom or one molecule. Therefore, how do we relate particles on an atomic and molecular level to the relatively large quantity of mass called a gram? We do it very simply by stating that 1 atomic mass unit will become 1 gram when we wish to compare masses of chemicals and carry out laboratory experiments. If we express the atomic mass of an element in grams, we call this quantity 1 mole. The term *mole* is commonplace in chemistry and is used to describe a certain number of molecules, atoms, ions, or electrons. The actual number of particles contained in a mole of any substance is called (1) _____ _____ and has a value of (2) _____. It is possible to speak about a mole of bricks or a mole of moles, but the word is usually restricted to a chemical meaning. It is important to remember that a mole of a chemical compound not only refers to Avogadro's number of formula units but also relates to a corresponding mass of these formula units which can be determined in the laboratory.

2. Calculate the molar mass for the following compounds with the aid of the short list of atomic masses.

			Atomic Masses (g/mol)
(1)	CCl_4	Na	22.99
(2)	$CaCO_3$	C	12.01
(3)	NH_4Cl	O	16.00
(4)	$NaClO$	Ca	40.08
(5)	$NaHCO_3$	Cl	35.45
		N	14.01
		H	1.008

Do your calculations here.

3. Using the molar masses that were calculated for the second question, find the number of moles present for the following masses of chemical. Be sure the units cancel out properly after you set up the problem.

   (1) 77 g CCl_4
   (molar mass = 153.8 g). Number of moles = _____

   (2) 33 g of $CaCO_3$
   (molar mass = 100.1 g). Number of moles = _____

   (3) 89 g NH_4Cl
   (molar mass = 53.49 g). Number of moles = _____

   Do your calculations here.

4. Using the molar masses that were calculated for the second question, find the number of grams represented by the following number of moles of chemical. Be sure to cancel out the units to make sure you have set the problem up correctly.

   (1) 1.5 moles of NaClO (molar mass = 74.44 g)   _____ g

   (2) 0.67 mole of $NaHCO_3$ (molar mass = 84.01 g)   _____ g

   Do your calculations here.

5. Using the partial list of atomic masses, calculate the number of moles represented by a certain mass of an element.

      Na   22.99 g/mol
      Ag   107.9 g/mol
      S   32.06 g/mol

(1) How many moles are represented by 46.0 g of Na atoms?

Do your calculations here.

(2) How many moles are represented by 54.0 g of Ag?

Do your calculations here.

(3) How many moles are represented by 100. g of S?

Do your calculations here.

6. Using the following list of molar masses, calculate the elemental percent composition for the four compounds listed.

   Molar masses   N = 14.01 g/mol,   O = 16.00 g/mol,   Al = 26.98 g/mol
   C = 12.01 g/mol,   Mg = 54.94 g/mol,   K = 39.10 g/mol

   (1) N_2O (2) Al_2O_3 (3) K_2CO_3 (4) $KMnO_4$

   Do your calculations here.

   (1) N_2O

   (2) Al_2O_3

   (3) K_2CO_3

   (4) $KMnO_4$

7. Fill in the blank space or circle the appropriate response.

The simplest formula of a compound, or the (1) _____ formula, tells us the (2) smallest/largest ratio of atoms present in a compound. The ratio is usually a small (3) _____ number ratio. It is possible for two compounds to have the same empirical formula, but different (4) _____ _____ formulas. The true formula for a compound, or (5) _____ formula, represents the actual number of (6) _____ of each element found in one molecule of a compound. The mass of all the (7) _____ in a molecular formula is the compound's (8) _____ .

8. From the percent composition data given for each compound, calculate the empirical formula.

(1)   86.6% Pb   Molar mass Pb = 207.2 g/mol
       13.4% S   Molar mass S = 32.06 g/mol
                                               Empirical formula _____

Do your calculations here.

(2)   27.1% Na   Molar mass Na = 22.99 g/mol
       16.5% N   Molar mass N = 14.01 g/mol
       56.4% O   Molar mass O = 16.00 g/mol
                                               Empirical formula _____

Do your calculations here.

9. A certain compound is found to contain 1.45 g of Na, 2.05 g of S, and 1.5 g of O. With the aid of the list of molar masses, calculate the empirical formula.

   Na   Molar mass   22.99 g/mol
   S   Molar mass   32.06 g/mol
   O   Molar mass   16.00 g/mol

                                       Formula _____

   Do your calculations here.

10. The molar mass of an unknown sugar substance is experimentally found to be 180. g/mol. The percent composition data are determined to be 40% carbon, 7.0% hydrogen, and 53% oxygen. Calculate first the empirical formula and then the molecular formula.

    C   Molar mass   12.01 g/mol
    H   Molar mass   1.008 g/mol
    O   Molar mass   16.00 g/mol

                                      Empirical _____

                                      Molecular _____

    Do your calculations here.

11. A compound of bromine and iodine is formed by direct reaction between the two elements. It is found that 5.40 g of bromine react with 8.58 g of iodine. What is the empirical formula for the compound and if the molar mass is 206.8 g/mol, what is the molecular formula? What is the percent composition of the compound?

    Do your calculations here.

12. Calculate the number of grams and the number of atoms represented by the following quantities of two elements.

    (1) 1.5 moles of Si
       Molar mass of Si = 28.09 g/mol

    (2) 0.30 mole of Ca
       Molar mass of Ca = 40.08 g/mol

    Do your calculations here.

Challenge Problem

13. A compound known as cadaverine (1,5-pentane diamine) is a ptomaine formed by the action of bacteria on meat and fish. Analysis shows that the elemental composition is C 58.8%, H 13.8% and N 27.4%. Determine the empirical formula and the molecular formula. ("Pentane" means 5 carbon atoms.)

    Do your calculations here.

14. How many grams of Fe contain the same number of atoms as 154 g of arsenic?

   Do your calculations here.

15. At the present time the population of the United States is approximately 265,000,000. If one mole of pennies were distributed among the entire population, how many dollars would each person receive?

   Do your calculations here.

RECAP SECTION

Chapter 7 puts you to work using the periodic table and chemical formulas to try out your math skills on some basic chemical problem solving. You will be interested to know that, prior to the late 1890s, many chemists spent their careers working on problems of the elemental composition of various compounds. The techniques and mathematical steps that they used are all related to the problems that you have just solved. Even today it is necessary to determine the empirical and molecular formulas of the many new compounds synthesized each year.

ANSWERS TO QUESTIONS AND SOLUTIONS TO PROBLEMS

1. (1) Avogadro's number (2) 6.022×10^{23}

2. (1) $12.01 \text{ g/mol} + (4 \times 35.45 \text{ g/mol}) = 153.8 \frac{g}{mol}$

   (2) $40.08 \text{ g/mol} + 12.01 \text{ g/mol} + (3 \times 16.00 \text{ g/mol}) = 100.1 \frac{g}{mol}$

   (3) $14.01 \text{ g/mol} + (4 \times 1.008 \text{ g/mol}) + 35.45 \text{ g/mol} = 53.49 \frac{g}{mol}$

   (4) $22.99 \text{ g/mol} + 35.45 \text{ g/mol} + 16.00 \text{ g/mol} = 74.44 \frac{g}{mol}$

   (5) $22.99 \text{ g/mol} + 1.008 \text{ g/mol} + 12.01 \text{ g/mol} + (3 \times 16.00 \text{ g/mol}) = 84.01 \frac{g}{mol}$

3. (1) molar mass of $CCl_4 = 153.8$ g
Therefore,
$$77 \text{ g} \times \frac{1 \text{ mol}}{153.8 \text{ g}} = 0.50 \text{ mol}$$

(2) molar mass of $CaCO_3 = 100.1$ g
Therefore,
$$33 \text{ g} \times \frac{1 \text{ mol}}{100.1 \text{ g}} = 0.33 \text{ mol}$$

(3) molar mass of $NH_4Cl = 53.49$ g
Therefore,
$$89 \text{ g} \times \frac{1 \text{ mol}}{53.49 \text{ g}} = 1.7 \text{ mol}$$

4. (1) 1.5 moles of NaClO will equal
$$1.5 \text{ mol} \times 74.44 \frac{\text{g}}{\text{mol}} = 1.1 \times 10^2 \text{ g}$$

(2) 0.67 mole of $NaHCO_3$ will equal
$$0.67 \text{ mol} \times 84.01 \frac{\text{g}}{\text{mol}} = 56 \text{ g}$$

5. (1) 2.00 moles of Na
You are asked how many moles are contained in 46.0 g of Na atoms. One mole of any element is equal to the atomic mass expressed in grams. For Na this amount is 22.99 g. What we have said is that 22.99 g of Na equals 1.0 mole. We have 46.0 g, which is more than 1.0 mole. Arranging the problem so that the units will cancel out properly and give us an answer that is greater than one, we have
$$46.0 \text{ g} \times \frac{1 \text{ mol}}{22.99 \text{ g}} = 2.00 \text{ mol}$$

(2) 0.500 mole
1 mole of Ag = 107.9 g

Therefore,
$$54.0 \text{ g} \times \frac{1 \text{ mol}}{107.9 \text{ g}} = 0.500 \text{ mol}$$

(3) 3.12 moles of S
The problem is solved just as the others have been.
$$100. \text{ g} \times \frac{1 \text{ mol}}{32.06 \text{ g}} = 3.12 \text{ mol S}$$

6. (1) N_2O

   molar mass $= (2 \times 14.01 \text{ g/mol}) + 16.00 \text{ g/mol} = 44.02 \dfrac{\text{g}}{\text{mol}}$

   Therefore,

   $\%N = \dfrac{2 \times 14.01 \text{ g/mol}}{44.02 \text{ g/mol}} \times 100 = 63.65\%$

   $\%O = \dfrac{16.00 \text{ g/mol}}{44.02 \text{ g/mol}} \times 100 = 36.35\%$

   (2) Al_2O_3

   molar mass $= (2 \times 26.98 \text{ g/mol}) + (3 \times 16.00 \text{ g/mol}) = 102.0 \dfrac{\text{g}}{\text{mol}}$

   Therefore,

   $\%Al = \dfrac{2 \times 26.98 \text{ g/mol}}{102.0 \text{ g/mol}} \times 100 = 52.90\%$

   $\%O = \dfrac{3 \times 16.00 \text{ g/mol}}{102.0 \text{ g/mol}} \times 100 = 47.06\%$

   (3) K_2CO_3

   molar mass $= (2 \times 39.10 \text{ g/mol}) + 12.01 \text{ g/mol} + (3 \times 16.00 \text{ g/mol}) = 138.2 \dfrac{\text{g}}{\text{mol}}$

   Therefore,

   $\%K = \dfrac{2 \times 39.10 \text{ g/mol}}{138.2 \text{ g/mol}} \times 100 = 56.58\%$

   $\%C = \dfrac{12.01 \text{ g/mol}}{138.2 \text{ g/mol}} \times 100 = 8.690\%$

   $\%O = \dfrac{3 \times 16.00 \text{ g/mol}}{138.2 \text{ g/mol}} \times 100 = 34.73\%$

   (4) $KMnO_4$

   molar mass $= 39.10 \text{ g/mol} + 54.94 \text{ g/mol} + (4 \times 16.00 \text{ g/mol}) = 158.0 \dfrac{\text{g}}{\text{mol}}$

   Therefore,

   $\%K = \dfrac{39.10 \text{ g}}{158.0 \text{ g}} \times 100 = 24.75\%$

   $\%Mn = \dfrac{54.94 \text{ g}}{158.0 \text{ g}} \times 100 = 34.77\%$

   $\%O = \dfrac{4 \times 16.00 \text{ g}}{158.0 \text{ g}} \times 100 = 40.51\%$

7. (1) empirical (2) smallest (3) whole (4) molecular
   (5) molecular (6) atoms   (7) atoms (8) molar mass

8. (1) PbS   (2) $NaNO_3$

   (1) Taking 100 g of the compound composed of Pb and S, we would have 87 g Pb and 13 g S. We need to determine the number of moles of Pb and S present in the 100 g. Therefore,

   $$Pb \quad 87 \text{ g} \times \frac{1 \text{ mol}}{207.2 \text{ g}} = 0.42 \text{ mol}$$

   $$S \quad 13 \text{ g} \times \frac{1 \text{ mol}}{32.06 \text{ g}} = 0.40 \text{ mol}$$

   The ratio of Pb to S is 0.42 to 0.40 or 1:1. The formula is PbS.

   (2) Likewise, in a 100 g of compound 2, we would have 27.1 g of Na, 16.5 g of N, and 56.4 g O. The number of moles of each element would be:

   $$Na \quad 27.1 \text{ g} \times \frac{1 \text{ mol}}{22.99 \text{ g}} = 1.18 \text{ mol}$$

   $$N \quad 16.5 \text{ g} \times \frac{1 \text{ mol}}{14.01 \text{ g}} = 1.18 \text{ mol}$$

   $$O \quad 56.4 \text{ g} \times \frac{1 \text{ mol}}{16.00 \text{ g}} = 3.53 \text{ mol}$$

   To eliminate the decimals from our ratio we must divide each of the numbers by the smallest number.

   $$Na = \frac{1.18}{1.18} = 1 \qquad O = \frac{3.53}{1.18} = 2.99$$

   $$N = \frac{1.18}{1.18} = 1$$

   When the ratio is expressed as small whole numbers, the correct empirical formula becomes $NaNO_3$.

9. $Na_2S_2O_3$

   We must determine the number of moles of each element present and then find the smallest whole-number ratio.

		Smallest Ratio
Na	$1.45 \text{ g} \times \frac{1 \text{ mol}}{22.99 \text{ g}} = 0.0631$ mole	$\frac{0.0631}{0.0631} = 1$
S	$2.05 \text{ g} \times \frac{1 \text{ mol}}{32.06 \text{ g}} = 0.0639$ mole	$\frac{0.0639}{0.0631} = 1$
O	$1.5 \text{ g} \times \frac{1 \text{ mol}}{16.00 \text{ g}} = 0.094$ mole	$\frac{0.094}{0.0631} = 1.5$

Therefore, $1:1:1.5 = 2:2:3$
The empirical formula is $Na_2S_2O_3$

10. Empirical formula: CH_2O Molecular formula: $C_6H_{12}O_6$

The first step is to find the empirical formula from the number of moles of each element for a hypothetical 100 g of compound.

Smallest Ratio

$$C \quad 40.g \times \frac{1 \text{ mol}}{12.01 \text{ g}} = 3.3 \text{ moles} \qquad \frac{3.3}{3.3} = 1$$

$$H \quad 7.0 \text{ g} \times \frac{1 \text{ mol}}{1.008 \text{ g}} = 6.9 \text{ moles} \qquad \frac{6.9}{3.3} = 2.1$$

$$O \quad 53 \text{ g} \times \frac{1 \text{ mol}}{16.00 \text{ g}} = 3.3 \text{ moles} \qquad \frac{3.3}{3.3} = 1$$

The rounded-off whole-number ratio would be $(CH_2O)_n$ and the empirical formula would therefore be CH_2O.

The molecular formula is calculated from the total atomic masses in the empirical formula and the molar mass. The total mass of the empirical formula is $12.01 + 2.016 + 16.00 = 30.03$. Next, determine the ratio between the given molar mass and the empirical formula mass.

$$\frac{180. \text{ g/mol}}{30.03 \text{ g/mol}} = 5.99$$

Therefore, we need to multiply the empirical formula by 6 to obtain the molecular formula.

$$6 \times CH_2O \text{ would be } C_6H_{12}O_6$$

11. Empirical formula and molecular formula are the same, BrI. The percent composition is 38.6% Br_2 and 61.3% I_2.

$$\text{Number of moles of } Br_2 = \frac{5.40 \text{ g}}{79.90 \text{ g/mol}} = 0.0676 \text{ mol}$$

$$\text{Number of moles of } I_2 = \frac{8.58 \text{ g}}{126.9 \text{ g/mol}} = 0.0676 \text{ mol}$$

The number of moles of each element are in a ratio of $1:1$, so the empirical formula is BrI. The mass of one mole of BrI is 206.8 g, which matches the value given in the problem, so the molecular formula is also BrI. Percent composition is calculated as follows:

$$\% \text{ Br} = \frac{\text{mass Br}_2}{\text{total mass}} = \frac{5.40 \text{ g}}{(5.40 + 8.58) \text{ g}} \times 100 = 38.6\%$$

Likewise

$$\% \text{ I}_2 = \frac{8.58 \text{ g}}{13.98 \text{ g}} \times 100 = 61.4\%$$

12. (1) 42 g of Si and 9.0×10^{23} atoms Si
    Silicon has 28.09 g in 1 mole. We have 1.5 moles, which will be more than 28.09 g. Arranging the problem so that the units cancel out, we have:

    $$1.5 \, \text{mol} \times 28.09 \, \frac{\text{g}}{\text{mol}} = 42 \text{ g of Si}$$

    To find the number of atoms in a certain number of moles, we simply multiply the number of moles by Avogadro's number.

    $$1.5 \, \text{mol} \times 6.022 \times 10^{23} \, \frac{\text{atoms}}{\text{mol}} = 9.0 \times 10^{23} \text{ atoms Si}$$

   (2) 12 g of Ca and 1.8×10^{23} atoms Ca

    $$0.30 \, \text{mol} \times 40.08 \, \frac{\text{g}}{\text{mol}} = 12 \text{ g of Ca}$$

    $$0.30 \, \text{mol} \times 6.022 \times 10^{23} \, \frac{\text{atoms}}{\text{mol}} = 1.8 \times 10^{23} \text{ atoms Ca}$$

13. The empirical and molecular formula are both $C_5H_{14}N_2$, since we know from the problem that the compound contains 5 C atoms.

    C $\dfrac{58.5 \text{ g}}{12.01 \text{ g/mol}} = 4.90$ mol

    N $\dfrac{27.4 \text{ g}}{14.01 \text{ g/mol}} = 1.96$ mol

    H $\dfrac{13.8 \text{ g}}{1.008 \text{ g/mol}} = 13.7$ mol

14. 115 g Fe. We can start by finding out how many moles of arsenic are represented by 154 g.

    $$\frac{154 \text{ g}}{74.92 \text{ g/mol}} = 2.06 \text{ mol As}$$

    2.06 moles of As contains the same number of atoms as 2.06 moles of Fe. To convert this value into grams of Fe, we need only to multiply the molar mass by the number of moles.

    $$2.06 \, \text{mol} \times 55.85 \, \frac{\text{g Fe}}{\text{mol}} = 115 \text{ g Fe (3 significant figures)}$$

15. $\$2.27 \times 10^{13}$ per person. One mole of pennies equals one Avogadro's number of pennies of 6.022×10^{23} pennies. To find dollars, divide this number by 10^2 or 100.

    $$\frac{6.022 \times 10^{23} \text{ pennies}}{1 \times 10^2 \, \frac{\text{pennies}}{\text{dollar}}} = 6.022 \times 10^{21} \text{ dollars}$$

    $$\frac{6.022 \times 10^{21} \text{ dollars}}{2.65 \times 10^8 \text{ persons}} = \$2.27 \times 10^{13}/\text{person}$$

CHAPTER EIGHT

Chemical Equations

SELF-EVALUATION SECTION

1. Match the symbols used in chemical equations with the corresponding descriptive statements

   Symbols

   $\rightarrow$ (s)

   $+$ (l)

   $\rightleftarrows$ (g)

        (Δ)

        (aq)

   (1) Gas (written after substance) _____

   (2) Reversible reaction; equilibrium between reactants and products _____

   (3) Heat _____

   (4) Added to _____

   (5) Liquid (written after substance) _____

   (6) Aqueous solution (substance dissolved in water) _____

   (7) Yields; produces (points to products) _____

   (8) Solid (written after substance) _____

2. Balance the following equations.

   (1) $Fe + H_2O \rightarrow Fe_3O_4 + H_2$

(2) $H_2O_2 \rightarrow H_2O + O_2$

(3) $NH_4NO_2 \rightarrow N_2 + H_2O$

(4) $C_6H_{14} + O_2 \rightarrow CO_2 + H_2O$

(5) $CO + Fe_3O_4 \rightarrow FeO + CO_2$

Translate the word equations into formulas and balance them.

(6) Potassium nitrate $\rightarrow$ Potassium nitrite + Oxygen

(7) Calcium oxide + Hydrochloric acid $\rightarrow$ Calcium chloride + Water

(8) Copper metal + Sulfuric acid $\rightarrow$ Copper (II) sulfate + Water + Sulfur dioxide

(9) Bromine + Hydrogen sulfide → Hydrogen bromide + sulfur

(10) Zinc sulfide + Oxygen → Zinc oxide + Sulfur dioxide

3. Balance the following equations

   (1) $K + H_2O \rightarrow KOH + H_2$

   (2) $Ca + O_2 \rightarrow CaO$

   (3) $Na_2O + H_2O \rightarrow NaOH$

   (4) $Zn + HCl \rightarrow ZnCl_2 + H_2$

   (5) $N_2 + H_2 \rightarrow NH_3$

Translate the word equations into formulas and balance them.

(6) Carbon + Oxygen → Carbon monoxide

(7) Sodium hydrogen carbonate + Sulfuric acid → Sodium sulfate + Water + Carbon dioxide

(8) Lithium hydroxide + Hydrobromic acid → Lithium bromide + Water

(9) Potassium Chromate + Lead (II) nitrate → Potassium nitrate + Lead (II) chromate

(10) Nitric acid + Calcium hydroxide → Calcium nitrate + Water

4. Identify the following reactions as combinations (C), decomposition (D), single displacement (SD), or double displacement (DD).

(1) $2\ Al(OH)_3 + 3\ H_2SO_4 \rightarrow Al_2(SO_4)_3 + 6\ H_2O$ _____

(2) $4\ K + O_2 \rightarrow 2\ K_2O$ _____

(3) $Cl_2 + 2\ NaBr \rightarrow Br_2 + 2\ NaCl$ _____

(4) $2\ HgO \rightarrow 2\ Hg + O_2$ _____

(5) $MgCl_2 + 2\ AgNO_3 \rightarrow 2\ AgCl + Mg(NO_3)_2$ _____

(6) $CaO + H_2O \rightarrow Ca(OH)_2$ _____

(7) $2\ HCl + Na_2CO_3 \rightarrow 2\ NaCl + H_2O + CO_2$ _____

(8) $2\ KClO_3 \rightarrow 2\ KCl + 3\ O_2$ _____

5. Identify the following reactions as exothermic (ex) or endothermic (en).

   (1) $N_2(g) + O_2(g) + 181 \text{ kJ} \rightarrow 2 \text{ NO}(g)$ _____

   (2) $C(s) + O_2(g) \rightarrow CO_2(g) + 94.0 \text{ kcal}$ _____

   (3) $C_3H_8(g) + 5 O_2(g) \rightarrow 3 CO_2(g) + 4 H_2O(g) + 2200 \text{ kJ}$ _____

6. Interpret the odd-numbered reactions of question 4 in terms of number of moles of reactants and products involved. Write your answers below.

   (1)

   (3)

   (5)

   (7)

Challenge Problem

7. Identify each of following reactions as combination, decomposition, single displacement or double displacement. Complete and balance each one.

   (1) $KClO_3 \xrightarrow{\Delta}$

   (2) $Na + H_2O \longrightarrow$

   (3) $HCl + K_2CO_3 \longrightarrow$

   (4) $H_2 + N_2 \xrightarrow{\Delta}$

   (5) $CaCO_3 \xrightarrow{\Delta}$

   (6) $Zn + Pb(NO_3)_2 \longrightarrow$

   (7) $Al_2(SO_4)_3 + NH_4OH \longrightarrow$

   (8) $CaO + H_2O \longrightarrow$

   (9) $Cl_2 + KBr \longrightarrow$

   (10) $Mg + NiCl_2 \longrightarrow$

   (11) $Ba(NO_3)_2 + Na_2SO_4 \longrightarrow$

(12) $NH_3 + HCl \longrightarrow$

(13) $H_2 + Cl_2 \longrightarrow$

(14) $HNO_3 + NaOH \longrightarrow$

(15) $NaCl + H_2SO_4 \longrightarrow$

(16) $Mg + H_2SO_4 \longrightarrow$

(17) $Na_2O + H_2SO_4 \longrightarrow$

(18) $Mg + O_2 \xrightarrow{\Delta}$

(19) $NaI + Cl_2 \longrightarrow$

(20) $AgNO_3 + NaCl \longrightarrow$

8. Will a reaction occur when the following are mixed? If so, write a balanced equation. Use the short activity series in the chapter.

(1) $Al(s) + NaBr(aq)$
(2) $Zn(s) + CuCl_2(aq)$
(3) $Cu(s) + HCl(aq)$
(4) $Mg(s) + NiCl(aq)$
(5) $K(s) + H_2O$

RECAP SECTION

Chapter 8 is a self-contained section similar to Chapter 6. You should refer to both of these chapters from time to time to review nomenclature and chemical equations. You have learned what a chemical equation is and how to put one together in a balanced form. We are now able to use a chemical equation rather than a word equation to describe chemical changes.

ANSWER TO QUESTIONS

1. (1) (g) (2) $\rightleftarrows$ (3) Δ (4) $+$
   (5) (l) (6) (aq) (7) $\rightarrow$ (8) (s)

2. (1) $3\ Fe + 4\ H_2O \rightarrow Fe_3O_4 + 4\ H_2$
   (2) $2\ H_2O_2 \rightarrow 2\ H_2O + O_2$
   (3) $NH_4NO_2 \rightarrow N_2 + 2\ H_2O$
   (4) $2\ C_6H_{14} + 19\ O_2 \rightarrow 12\ CO_2 + 14\ H_2O$
   (5) $CO + Fe_3O_4 \rightarrow 3\ FeO + CO_2$
   (6) $2\ KNO_3 \rightarrow 2\ KNO_2 + O_2$
   (7) $CaO + 2\ HCl \rightarrow CaCl_2 + H_2O$
   (8) $Cu + 2\ H_2SO_4 \rightarrow CuSO_4 + 2\ H_2O + SO_2$
   (9) $Br_2 + H_2S \rightarrow 2\ HBr + S$
   (10) $2\ ZnS + 3\ O_2 \rightarrow 2\ ZnO + 2\ SO_2$

3. (1) $2\ K + 2\ H_2O \rightarrow 2\ KOH + H_2$
    (2) $2\ Ca + O_2 \rightarrow 2\ CaO$
    (3) $Na_2O + H_2O \rightarrow 2\ NaOH$
    (4) $Zn + 2\ HCl \rightarrow ZnCl_2 + H_2$
    (5) $N_2 + 3\ H_2 \rightarrow 2\ NH_3$
    (6) $2\ C + O_2 \rightarrow 2\ CO$
    (7) $2\ NaHCO_3 + H_2SO_4 \rightarrow Na_2SO_4 + 2\ H_2O + 2\ CO_2$
    (8) $LiOH + HBr \rightarrow LiBr + H_2O$
    (9) $K_2CrO_4 + Pb(NO_3)_2 \rightarrow 2\ KNO_3 + PbCrO_4$
    (10) $2\ HNO_3 + Ca(OH)_2 \rightarrow Ca(NO_3)_2 + 2\ H_2O$

4. (1) DD   (2) C   (3) SD   (4) D
    (5) DD   (6) C   (7) DD   (8) D

5. (1) en   (2) ex   (3) ex

6. (1) Reactants 2 moles $Al(OH)_3$ and 3 moles H_2SO_4
        Products   1 mole $Al_2(SO_4)_3$ and 6 moles H_2O

    (3) Reactants 1 mole Cl_2 and 2 moles $NaBr$
        Products   1 mole Br_2 and 2 moles $NaCl$

    (5) Reactants 1 mole $MgCl_2$ and 2 moles $AgNO_3$
        Products   2 moles $AgCl$ and 1 mole $Mg(NO_3)_2$

    (7) Reactants 2 moles HCl and 1 mole Na_2CO_3
        Products   2 moles $NaCl$ and 1 mole H_2O and 1 mole CO_2

7. (1) decomposition $2\ KClO_3 \xrightarrow{\Delta} 2\ KCl + 3\ O_2(g)$

    (2) single displacement $2\ Na + 2\ H_2O \rightarrow 2\ NaOH + H_2(g)$

    (3) double displacement $2\ HCl + K_2CO_3 \rightarrow 2\ KCl + H_2O + CO_2(g)$

    (4) combination $3\ H_2 + N_2 \xrightarrow{\Delta} 2\ NH_3$

    (5) decomposition $CaCO_3 \xrightarrow{\Delta} CaO + CO_2(g)$

    (6) single displacement $Zn + Pb(NO_3)_2 \rightarrow Zn(NO_3)_2 + Pb$

    (7) double displacement $Al_2(SO_4)_3 + 6\ NH_4OH \rightarrow 3\ (NH_4)_2SO_4 + 2\ Al(OH)_3$

    (8) combination $CaO + H_2O \rightarrow Ca(OH)_2$

    (9) single displacement $Cl_2 + 2\ KBr \rightarrow Br_2 + 2\ KCl$

    (10) single displacement $Mg + NiCl_2 \rightarrow Ni + MgCl_2$

    (11) double displacement $Ba(NO_3)_2 + Na_2SO_4 \rightarrow BaSO_4 + 2\ NaNO_3$

(12) combination $NH_3 + HCl \rightarrow NH_4Cl$

(13) combination $H_2 + Cl_2 \rightarrow 2\ HCl$

(14) double displacement $HNO_3 + NaOH \rightarrow NaNO_3 + H_2O$

(15) double displacement $NaCl + H_2SO_4 \rightarrow NaHSO_4 + HCl(g)$

(16) single displacement $Mg + H_2SO_4 \rightarrow MgSO_4 + H_2(g)$

(17) double displacement $Na_2O + H_2SO_4 \rightarrow Na_2SO_4 + H_2O$

(18) combination $2\ Mg + O_2 \xrightarrow{\Delta} 2\ MgO$

(19) single displacement $2\ NaI + Cl_2 \rightarrow 2\ NaCl + I_2$

(20) double displacement $AgNO_3 + NaCl \rightarrow AgCl + NaNO_3$

8. (1) No

(2) Yes $Zn(s) + CuCl_2(aq) \rightarrow Cu(s) + ZnCl_2(aq)$

(3) No

(4) Yes $Mg(s) + NiCl_2(aq) \rightarrow Ni(s) + MgCl_2(aq)$

(5) Yes $2\ K(s) + 2\ H_2O \rightarrow 2\ KOH + H_2(g)$

CHAPTER NINE

Calculations from Chemical Equations

SELF-EVALUATION SECTION

1. Fill in with the appropriate response.

    $$2\ Na + Cl_2 \rightarrow 2\ NaCl$$

    In the above equation, (1) _____ mole(s) Na react with (2) _____ mole(s) Cl_2 to give (3) _____ mole(s) NaCl. The mole ratio of Na to Cl_2 is (4) _____ . The mole ratio of Cl_2 to NaCl is (5) _____ . If 7.0 moles of Na react, (6) _____ mole(s) of Cl_2 react with it to produce (7) _____ mole(s) of NaCl.

    $$Pt + 8\ HCl + 2\ HNO_3 \rightarrow H_2PtCl_6 + 2\ NOCl + 4\ H_2O$$

    Chloroplatinic   Nitrosyl
Acid   Chloride

    In the above equation, (8) _____ mole(s) of Pt reacts with (9) _____ mole(s) HCl and (10) _____ mole(s) HNO_3 to give (11) _____ mole(s) of H_2PtCl_6, (12) _____ mole(s) NOCl and (13) _____ mol(s) H_2O. The mole ratio of Pt to HCl is (14) _____ ; the mole ratio of HCl to H_2PtCl_6 is (15) _____ . If 0.300 mole of Pt is reacted, (16) _____ mole(s) HCl and (17) _____ mole(s) of HNO_3 will react with the Pt and (18) _____ mole(s) of H_2PtCl_6 or (19) _____ grams of H_2PtCl_6 will be produced.

Do your calculations here.

2. $2 C_6H_{14} + 19 O_2 \rightarrow 12 CO_2 + 14 H_2O$

In the above equation, the mole ratio of C_6H_{14} to CO_2 is (1) _____, and the mole ratio of C_6H_{14} to H_2O is (2) _____. If oxygen is present in abundance to carry out the reaction, which reagent will be the limiting reactant? (3) _____. How many moles of CO_2 can be produced from 3 moles of C_6H_{14}? (4) _____. If only 1.0 mole of C_6H_{14} is available for the reaction, what is the theoretical yield of H_2O in moles and in grams? (5) _____ ; (6) _____. For the same amount of C_6H_{14} (1 mole), what is the theoretical yield of H_2O in moles and in grams? (7) _____ ; (8) _____. After carrying out the above reaction with 1.0 mole of C_6H_{14}, a chemist measured an actual yield of 210. g of CO_2 and 115. g of water. What was her percent yield of CO_2? (9) _____. Her percent yield of H_2O? (10) _____.

Do your calculations here.

3. We will now use the techniques of Chapter 9 to gain further chemical equation problem-solving skill.

Balance the following equation and then calculate the requested quantities.

$$PbO_2 \xrightarrow{\Delta} PbO + O_2(g)$$

(1) How many grams of O_2 can be obtained from 100. grams of PbO_2? This is the theoretical yield. Remember that we need to (a) use a balanced equation, (b) determine the number of moles of starting substance, (c) calculate the number of moles of desired substance using the mole-ratio technique, and (d) convert moles to grams of desired substance.

(2) What is the percent yield if the actual yield of O_2 in the above reaction was 5.0 grams?

Do your calculations here.

4. Very often in industrial chemical processes, one of the reactants will be present in an amount that exceeds the requirements of the balanced equation. The reactant that is not in excess will therefore limit the amount of product that is formed and is named the limiting reactant. Using the equation given, answer the questions below.

$$2\ Al(OH)_3 + 3\ H_2SO_4 \rightarrow Al_2(SO_4)_3 + 6\ H_2O$$

(1) The reaction is run with excess of sulfuric acid. What is the limiting reactant?
(2) You have 9 moles of H_2SO_4 present for the reaction. How many moles of $Al(OH)_3$ are required to react completely with this amount of H_2SO_4?
(3) If 4 moles of $Al(OH)_3$ is the amount available, how much of the 9 moles of H_2SO_4 can be used?
(4) Using the 4 moles of $Al(OH)_3$, how many moles of $Al_2(SO_4)_3$ and H_2O will you be able to produce?

Do your calculations here.

5. In the following reaction, how many moles of ZnO can be obtained? Also determine which reactant is the limiting reactant and which reactant is in excess.

$$2\ ZnS + 3\ O_2 \rightarrow 2\ ZnO + 2\ SO_2$$
100. g   100. g

Do your calculations here.

6. Balance the following equation and calculate how many moles of Cu can be formed from 5.0 moles of Al and 10.0 moles of $CuSO_4$. What is the limiting reactant and how much of the excess reactant is left after the reaction?

$$Al + CuSO_4 \rightarrow Cu + Al_2(SO_4)_3$$
   5.0 mol   10.0 mol

Do your calculations here.

7. What is the theoretical yield of Fe_2O_3 that can be produced from 3.0 kg of Fe according to the following unbalanced equation?

$$Fe + O_2 \rightarrow Fe_2O_3$$

Do your calculations here.

Challenge Problems

8. It is possible to reclaim silver from used photographic fixer by using the active metal, powdered zinc. Zinc replaces the silver in solution, followed by conversion of the silver to silver oxide. After filtering, the silver oxide is reduced to metallic silver by carbon. The series of reactions are:

   (1) $2\ AgBr + Zn \longrightarrow ZnBr_2 + 2\ Ag$

   (2) $4\ Ag + O_2 \longrightarrow 2\ Ag_2O$

   (3) $2\ Ag_2O + C \longrightarrow CO_2 + 4\ Ag$

How much silver can be reclaimed from 2.000 gallons of used fixer (density 1.018 g/mL) if the silver concentration is 300. parts per million? Part per million is a general purpose concentration term that can take on a variety of units. For example 1 µg per gram, 1 µL per liter and 1 mg per kilogram are examples of 1 ppm concentration.

Do your calculations here.

9. Refer to the problem above and the last reaction. The carbon that reduces the Ag_2O to metallic Ag comes from the filter paper that was used in the filtration step following reaction (2). Assuming the filter paper to be pure cellulose with an empirical formula of $C_6H_{12}O_6$, what mass of filter paper is required to reduce the 2.31 g of silver contained in the two gallon volume of fixer?

Do your calculations here.

RECAP SECTION

The principal objective of Chapter 9 is to present a logical method for attacking problem solving associated with chemical equations. Since chemistry deals with chemical reactions and reactions are expressed in equation form, it follows that calculations associated with reactions are about as relevant and practical as any topic in chemistry. Professionals in agriculture, home economics, forestry, biology, and chemical engineering, in addition to chemists, are constantly working with stoichiometric calculations.

The technique of using the mole-ratio method for stoichiometric problems is very important to learn. Once you have a balanced equation, it is possible to set up any ratio between two species in the equation. If the problem asks for grams of reactant to produce so many grams of product, there will be additional calculation steps to perform on each side of the ratio, but the ratio is the connecting link.

grams reactant A → moles reactant A → moles product B → grams product B

ANSWERS TO QUESTIONS AND SOLUTIONS TO PROBLEMS

1. (1) 2   (2) 1   (3) 2   (4) 2:1   (5) 1:2
   (6) 3.5   (7) 7   (8) 1   (9) 8   (10) 2
   (11) 1   (12) 2   (13) 4   (14) 1:8   (15) 18:1
   (16) 2.40   (17) 0.600   (18) 0.300
   (19) 0.300 mole times the molar mass will give us the number of grams produced by the reaction. The molar mass of

   $$H_2PtCl_6 = (2 \times 1.008 \text{ g/mol}) + 195.1 \text{ g/mol} + (6 \times 35.45 \text{ g/mol}) = 409.8 \frac{g}{mol}$$

   $$\text{Number of grams} = 409.8 \frac{g}{mol} \times 0.300 \text{ mol}$$

   $$= 123 \text{ g } H_2PtCl_6 \text{ (3 significant figures)}$$

2. (1) 2:12 or 1:6   (2) 2:14 or 1:7   (3) C_6H_{14}
   (4) 18   (5) 6
   (6) 6 moles times the molar mass will give us the number of grams of CO_2 produced by the reaction.

   The molar mass of $CO_2 = 12.01 \text{ g/mol} = (2 \times 16.00 \text{ g/mol}) = 44.01 \frac{g}{mol}$

   $$\text{Number of grams} = 44.01 \frac{g}{mol} \times 6 \text{ mol} = 264.1 \text{ g } CO_2$$

   (7) 7 moles
   (8) 7 moles times the molar mass will give us the number of grams of H_2O produced by the reaction.

   The molar mass of $H_2O = (2 \times 1.008 \text{ g/mol}) + 16.00 \text{ g/mol} = 18.02 \frac{g}{mol}$

   $$\text{Number of grams} = 18.02 \frac{g}{mol} \times 7 \text{ mol} = 126.1 \text{ g } H_2O$$

   (9) Percent yield of

   $$CO_2 = \frac{\text{actual yield}}{\text{theoretical yield}} \times 100$$
   $$= \frac{210. \text{ g}}{264 \text{ g}} \times 100$$
   $$= 79.5\%$$

   (10) Percent yield of

   $$H_2O = \frac{\text{actual yield}}{\text{theoretical yield}} \times 100$$
   $$= \frac{115 \text{ g}}{126 \text{ g}} \times 100$$
   $$= 91.3\%$$

3. (1) 6.69 g of O_2
   The equation must be balanced before we can determine a proper mole ratio for calculating the amount of O_2.

   $$2\ PbO_2 \xrightarrow{\Delta} 2\ PbO + O_2(g)$$

   Next, we need to convert 100. grams of PbO_2 into the number of moles of PbO_2.

   The molar mass of $PbO_2 = 207.2$ g/mol $+ (2 \times 16.00$ g/mol$) = 239.2\ \frac{g}{mol}$

   Number of moles of $PbO_2 = 100.\ g \times \frac{1\ mol}{239.2\ g} = 0.418$ mol

   The mole ratio is $\frac{mol\ desired\ substance}{mol\ starting\ substance} = \frac{1\ mol\ O_2}{2\ mol\ PbO_2}$

   moles of $O_2 = 0.418$ mol $PbO_2 \times \frac{1\ mol\ O_2}{2\ mol\ PbO_2} = 0.209$ mol O_2

   To convert moles of O_2 into grams, we multiply the number of moles by the molar mass.

   0.209 mol $\times 32.00\ \frac{g}{mol} = 6.69$ g O_2

   (2) 75%
   The percent yield is determined by dividing the actual yield by the theoretical yield multiplied by 100.

   $$\frac{actual}{theoretical} \times 100$$

   $$\frac{5.0\ g}{6.7\ g} \times 100 = 75\%$$

4. (1) $Al(OH)_3$, aluminum hydroxide   (2) 6 moles   (3) 6 moles
   (4) 2 moles $Al_2(SO_4)_3$, 12 moles H_2O

5. 1.03 mole ZnO, ZnS is limiting reactant, O_2 is in excess. First we need to determine the number of moles of ZnO that can be obtained from each of the reactants.

   $$100.\ g\ ZnS \times \frac{1\ mol\ ZnS}{97.44\ g\ ZnS} \times \frac{2\ mol\ ZnO}{2\ mol\ ZnS} = 1.03\ mol\ ZnO$$

   $$100.\ g\ O_2 \times \frac{1\ mol\ O_2}{32.00\ g\ O_2} \times \frac{2\ mol\ ZnO}{3\ mol\ O_2} = 2.08\ mol\ ZnO$$

6. The balanced equation is

   $$2\ Al + 3\ CuSO_4 \longrightarrow 3\ Cu + Al_2(SO_4)_3$$

   First we need to determine the number of moles of Cu that can be formed from each reactant.

   $$5.0\ mol\ Al \times \frac{3\ mol\ Cu}{2\ mol\ Al} = 7.5\ mol\ Cu$$

$$10.0 \text{ mol CuSO}_4 \times \frac{3 \text{ mol Cu}}{3 \text{ mol CuSO}_4} = 10.0 \text{ mol Cu}$$

Therefore, Al is the limiting reactant and 7.5 mol of Cu can be formed. Next we need to calculate the number of moles of $CuSO_4$ that will react with 5.0 moles of Al.

$$5.0 \text{ mol Al} \times \frac{3 \text{ mol CuSO}_4}{2 \text{ mol Al}} = 7.5 \text{ mol CuSO}_4$$

Therefore, $10.0 \text{ mol CuSO}_4 - 7.5 \text{ mol CuSO}_4 = 2.5 \text{ mol of CuSO}_4$ in excess.

7. 4300 g Fe_2O_3 (2 significant figures)

   The balanced equations is

   $$4 \text{ Fe} + 3 \text{ O}_2 \rightarrow 2 \text{ Fe}_2\text{O}_3$$

   The number of moles of $\text{Fe} = 3.0 \text{ kg} \times 1000 \frac{g}{kg} \times \frac{1 \text{ mol}}{55.85 \text{ g}}$

   $$= 54 \text{ mol of Fe}$$

   $$\text{moles of Fe}_2\text{O}_3 = 54 \text{ mol Fe} \times \frac{2 \text{ mol Fe}_2\text{O}_3}{4 \text{ mol Fe}} = 27 \text{ mol}$$

   To determine the number of grams, we multiply 27 moles of Fe_2O_3 by the molar mass.

   The molar mass of $Fe_2O_3 = (2 \times 55.85 \text{ g/mol}) + (3 \times 16.00 \text{ g/mol}) = 159.7 \frac{g}{mol}$

   $$27 \text{ mol} \times 159.7 \frac{g}{mol} = 4300 \text{ g Fe}_2\text{O}_3 \text{ (2 significant figures)}$$

8. 2.31 g

   2.000 gallons is equal to 7568 mL which has a density of 1.018 g/mL. The mass of the solution therefore is:

   $$(1.018 \text{ g/mL})(7568 \text{ mL}) = 7704 \text{ g}$$

   Of this mass, 7704 g, 300. ppm are silver ion. To calculate how many grams of silver might be reclaimed with no loss, we set up the following equation.

   $$\left(\frac{1}{1 \times 10^6 \text{ ppm}}\right)(300. \text{ ppm})(7704 \text{ g}) = 2.31 \text{ g (3 significant figures)}$$

9. 0.144 g filter paper

   The equation tells us that 1 mole of C reduces 4 moles of Ag ion to metallic silver.

   Therefore, moles of carbon required is equal to

   $$\frac{2.31 \text{ g Ag}}{107.9 \text{ g/mol}} \times \frac{1 \text{ mol C}}{4 \text{ mol Ag}} = 0.00535 \text{ mol carbon}$$

   We need a piece of filter paper that will contain at least 0.00535 moles carbon. The percentage of carbon in the cellulose is

   $$\%C = \frac{72.0 \text{ g}}{162. \text{ g}} \times 100 = 44.4\%$$

To calculate the mass of filter paper needed we can convert the number of moles of carbon into grams and divide this value by the amount of carbon in the paper.

$$\text{paper mass} = (0.00535 \text{ mol C})(12.0 \text{ g/mol})\left(\frac{1}{0.444}\right) = 0.144 \text{ g}$$

CHAPTER TEN

Modern Atomic Theory and the Periodic Table

SELF-EVALUATION SECTION

1. Circle the appropriate response.

Niels Bohr, a Danish physicist, suggested that the electron for the simplest atom, hydrogen, didn't wander randomly about the nucleus, but moved only in certain well-defined orbits. You may have seen an atomic symbol used by a power company or other agency illustrating this model, which is analogous to orbits of planets around the sun. Bohr's theory states that there are various well-defined orbits available to an electron, some close to the nucleus and some farther away. When an electron absorbs energy, it jumps to a (1) higher/lower energy level, and as it loses energy, it moves to a (2) higher/lower energy level; the excess energy is given off as light energy. Bohr's theory explains very nicely the atomic spectra of hydrogen, but it does not explain spectra of more complicated multi-electron atoms. The quantum mechanic theory has replaced Bohr's theory today as the best explanation for electron behavior. The significant difference in the two theories for us is that quantum mechanics states that it is impossible to know exactly the position of an electron at any instant in time. What we can say and visually represent is that electrons occupy orbitals around the nucleus which are regions of space where the electrons most probably are. We often use the term *electron cloud* to describe the region of space defined by the various energy levels. The word *cloud* is appropriate because it pictures the somewhat nebulous behavior of an electron. Quantum or wave mechanic equations provide us with a tool to describe the probable location of electrons in an atom.

2. Write the electron configuration for the elements listed, using the basic rules and energy level order regarding the state of electrons in atoms listed as a, b, and c.
    a. In the ground state (lowest energy state) of an atom, electrons tend to occupy orbitals of the lowest possible energy.
    b. Each orbital may contain a maximum of two electrons (with opposite spins).
    c. The energy level order is.

    $1s, 2s, 2p, 3s, 4s, 3d$

The maximum number of electrons that can be found in any main energy level can be determined by a formula involving the term n where n is the (1) _____.
When $n = 1$, the total number of electrons will be (2) _____, which will fill the sublevels (3) *s, p, d, f*. For $n = 2$, the total number of electrons will be (4) _____, which will fill the sublevels (5) *s, p, d, f*. Look at the following symbol and describe its meaning:

$4s^2$

The number 4 represents (6) _____, the small letter s represnts (7) _____, and the superscript 2 represents (8) _____.

Now we can write electron configurations. For example, lithium has atomic number 3. Therefore, the correct way of writing the electron configuration would be Li $1s^2 2s^1$.

Try the configurations for
(9) Nitrogen   Atomic number 7
(10) Argon   Atomic number 18
(11) Sodium   Atomic number 11
(12) Iron   Atomic number 26
(13) Calcium Atomic number 20

Write your answers here.

Determine the atomic number for the element represented by the following electron configurations.

(14) $1s^2 2s^2 2p^6 3s^2 3p^2$ _____
(15) $1s^2 2s^2 2p^4$ _____
(16) $1s^2 2s^2 2p^6 3s^2 3p^6 4s^2 3d^7$ _____
(17) $1s^2 2s^2 2p^6 3s^2 3p^6 4s^2 3d^1$ _____
(18) $1s^2 2s^2 2p^6 3s^2 3p^6 4s^2$ _____

3. Atomic structures may also be represented by diagrams as shown on page 203 of the text. Using this method, diagram the electron structure for:

   (1) $_6C$
   (2) $_{19}K$
   (3) $_{10}Ne$
   (4) $_{23}V$
   (5) $_{33}As$
   (6) $_{17}Cl$

   Write your answers here.

4. Fill in the blank space or circle the appropriate response.

   The periodic table, which is a very useful source of information for chemists and other scientists, arranges elements in a certain way based on electronic structure. The word *periodicity* suggests that elemental properties in the periodic table occur (1) only once/repeatedly . If the known elements are arranged in a table according to increasing atomic number rather than increasing atomic (2) _____ , we can begin to talk about various groupings of elements. A horizontal group of elements in the periodic table is called a (3) _____ . A vertical group of elements is called a (4) _____ . Each period of elements is numbered with an integer that indicates the outermost (5) _____ being filled progressively with electrons. The members of a group or (6) _____ of elements are related to each other in that the outermost energy level of each contains (7) the same/a different number of electrons. Because of the electron configurations, groups and families of elements will have (8) similar/different chemical properties. Beginning with the 4th period, (9) terminal/transition elements occur in the (10) middle/end of the period. These elements are all (11) nonmetals/metals and form a variety of compounds. The distinguishing characteristic of the transition elements is that an (12) inner/outer shell of electrons is filled as the atomic number increases. The distinguishing electrons for the transition elements are (13) s & p/d & f .

Focusing our attention on the various groups or families of elements, we can make some general statements concerning group characteristics. For Groups IA, VIIIA, IB, and IIB, the group number is the same as the number of (14) _____ in the outer energy level. The other groups of elements, the transition metals, are filling (15) inner/outer electronic orbitals and cannot be classified as representative elements. In addition, the noble gases (with the exception of helium) have (16) _____ electrons in their outer energy level.

The elements on the left side of the periodic table are (17) metals/nonmetals and share the general characteristics of all metals. The elements at the top of the groups on the right side of the periodic table are (18) metals/nonmetals. However, as one proceeds down these groups toward the elements with larger nuclei, the metallic characteristics of the elements (19) increase/decrease. Thus, Group VA begins with nitrogen, a nonmetal, and ends with bismuth, a metal. In general, as one goes across a period the atomic radii (20) decrease/increase as the atomic number increases.

Since all elements in a group have (21) different/the same valence structure, one can expect that elements in a group will have (22) similar/varied chemical properties. For example, if magnesium (Mg) forms a compound with oxygen with the formula MgO, then a compound formed from barium (Ba), which is in the same group as magnesium, and oxygen, would have the formula (23) _____ . In another example, aluminum oxide has a formula of Al_2O_3. Sulfur, which is in the same group as oxygen, forms a compound with aluminum named aluminum sulfide, which has the formula (25) _____ .

5. Using the periodic table in Chapter 10 of the textbook, find the following items of information about each element.

		Symbol	Atomic Number	Atomic Mass	No. of Outer Orbital Electrons
(1)	Phosphorus				
(2)	Fluorine				
(3)	Mercury				
(4)	Cesium				

Challenge Problems

6. After reviewing examples in the text for a few minutes, try your hand at constructing the orbital diagram ($1s$, $2s$, etc.) for (1) $_{21}Sc$ (2) $_{29}Cu$ (3) $_{40}Zr$.

   Write your answers here.

7. Mendeleev, in formulating his periodic arrangement of the elements, was able to predict in 1871 the properties of two as yet undiscovered elements: eka-silicon and eka-aluminum. Using reference sources such as the Handbook of Chemistry and Physics (Chemical Rubber Co.), and a periodic table, fill in the following table and answer the questions.

	eka-Aluminum	eka-Silicon
Atomic mass predicted	70.89	73.40
actual	_____	_____
Specific gravity predicted	5.9	5.5
actual	_____	_____
Specific heat predicted	0.381 J/g°C	0.305 J/g°C
actual	_____	_____
Identification of element	_____	_____

— 83 —

RECAP SECTION

In Chapter 10 the model of the atom is modified to our modern ideas. The Bohr model made the first attempt to locate the electrons within the atom. Unfortunately, the concept of orbits could not be used beyond hydrogen and a more accurate model was soon devised. In modern theory of the atom the actual path of the electron is not known. Instead, electrons occupy regions of space known as orbitals. This results in an atomic model of a small dense nucleus, surrounded by a series of electron clouds. The knowledge of energy levels and sublevels allows us to write electron configuration. The special stability of the noble gases can be attributed to the stability associated with 8 electrons in the valence electron level (i.e, the s and p orbitals of the outer energy level are filled). The material in Chapter 10 has also introduced us to a most important practical tool, the periodic table. The information contained in the complete long form of the periodic table is used by scientists everywhere during their day-to-day work. As you continue on in science, you will find yourself referring to the table to calculate molar masses of chemicals, to look up physical properties, and to determine theoretically possible chemical formulas.

ANSWERS TO QUESTIONS AND SOLUTIONS TO PROBLEMS

1. (1) higher (2) lower

2. (1) principal energy level
   (2) 2 (3) s (4) 8 (5) s, p (6) principal energy level
   (7) Type of sublevel (8) Number of electrons in that sublevel
   (9) $1s^2 2s^2 2p^3$ (10) $1s^2 2s^2 2p^6 3s^2 3p^6$ (11) $1s^2 2s^2 2p^6 3s^1$
   (12) $1s^2 2s^2 2p^6 3s^2 3p^6 4s^2 3d^6$ (13) $1s^2 2s^2 2p^6 3s^2 3p^6 4s^2$
   (14) 14 (15) 8 (16) 27 (17) 21 (18) 20

3. (1) C

	1s	2s	2p		
	↑↓	↑↓	↑	↑	

   (2) K

	1s	2s	2p			3s	3p			4s
	↑↓	↑↓	↑↓	↑↓	↑↓	↑↓	↑↓	↑↓	↑↓	↑

   (3) Ne

	1s	2s	2p		
	↑↓	↑↓	↑↓	↑↓	↑↓

   (4) V

	1s	2s	2p			3s	3p			4s
	↑↓	↑↓	↑↓	↑↓	↑↓	↑↓	↑↓	↑↓	↑↓	↑↓

   3d

↑	↑			

(5) As

	1s	2s	2p			3s	3p			4s
	↑↓	↑↓	↑↓	↑↓	↑↓	↑↓	↑↓	↑↓	↑↓	↑↓

	3d					4p		
	↑↓	↑↓	↑↓	↑↓	↑↓	↑	↑	↑

(6) Cl

	1s	2s	2p			3s	3p		
	↑↓	↑↓	↑↓	↑↓	↑↓	↑↓	↑↓	↑↓	↑

4. (1) repeatedly   (2) mass   (3) period   (4) group or family
    (5) energy level   (6) family   (7) the same   (8) similar
    (9) transition   (10) middle   (11) metals   (12) inner
    (13) d & f   (14) valence electrons (15) inner   (16) eight
    (17) metals   (18) nonmetals   (19) increase   (20) decrease
    (21) the same   (22) similar   (23) BaO   (24) Al$_2$O$_3$

5. (1) P   15   30.97   5
    (2) F   9   19.00   7
    (3) Hg   80   200.6   2
    (4) Cs   55   132.9   1

6. (1) $_{21}$Sc $1s^2\ 2s^2\ 2p^6\ 3s^2\ 3p^6\ 4s^2\ 3d^1$
    (2) $_{29}$Cu $1s^2\ 2s^2\ 2p^6\ 3s^2\ 3p^6\ 4s^2\ 3d^{10}$
    (3) $_{40}$Zr $1s^2\ 2s^2\ 2p^6\ 3s^2\ 3p^6\ 4s^2\ 3d^{10}\ 4p^6\ 5s^2\ 4d^2$

7.

	eka-Aluminum	eka-Silicon
Atomic mass actual	69.72	72.59
Specific gravity	5.9	5.3
Specific heat	0.368 J/g°C	0.322 J/g°C
Identification	Gallium	Germanium

CHAPTER ELEVEN
Chemical Bonds – The Formation of Compounds from Atoms

SELF-EVALUATION SECTION

1. Fill in the blank space or circle the appropriate response.

   When a neutral atom loses an orbital electron, it becomes a (1) <u>positively/ negatively</u> charged ion. The energy required to remove a mole of electrons from a mole of atoms is called the (2) _____ energy and is a relatively (3) <u>low/high</u> value for Group I and Group II metals and (4) <u>low/high</u> for non-metallic elements. The ionization energy for the noble gas elements is especially (5) <u>high/low</u>, indicating that eight electrons in the valence shell of an atom is a very stable structure. If you list the ionization energies of the elements from a group in the periodic table, the element from the top of the group has a (6) <u>higher/lower</u> ionization energy than the element at the bottom of the group. Two factors account for this experimental observation. As you go down a group, the electron being removed is (7) <u>farther from/closer to</u> the nucleus, and the increasing number of filled electron orbitals shields the valence electrons from the positive nucleus. The valence electrons are shown in the Lewis structures. We will use Lewis structures in the next section to illustrate the concept of forming chemical bonds.

   As mentioned above, when a neutral atom (8) <u>loses/gains</u> an electron it becomes positively charged. A positively charged ion is also called a (9) _____. For example, a potassium atom has 19 protons and 19 electrons. A potassium ion has a charge of +1, which means the ion has (10) _____ electrons instead of 19, as in the neutral atom. Conversely, a negatively charged ion, which is called an (11) _____, is formed when a neutral atom (12) <u>loses/gains</u> electrons. A chlorine atom has 17 protons and 17 electrons and becomes an anion by (13) <u>gaining/losing</u> an electron. A chloride ion has a stable outer shell of (14) _____ electrons. A sodium atom in close proximity to a chlorine atom can also reach a stable outer shell of eight electrons by losing one electron. Thus, both sodium and chlorine reach a stable electron structure by the process of electron transfer. The metallic elements attain a stable structure by (16) <u>gaining/losing</u> electrons; the nonmetallic elements attain a stable structure by (17) <u>gaining/losing</u> electrons.

2. In this section, we will use Lewis structures to illustrate how chemical bonds are formed. Using the techniques in Sections 11.4 and 11.5, we will work out some electron-transfer problems, then some dealing with electron sharing.

   (1) Using Lewis structures, show how the compound zinc iodide (ZnI_2) is formed. Zinc is a Group IIB element and iodine is a Group VIIA element.

   (2) Show how the compound potassium oxide (K_2O) is formed. Potassium is a Group IA element, and oxygen is a Group VIA element.

   (3) Let's try some problems that involve electron sharing between atoms to obtain a stable configuration of eight electrons. Show how the compound phosphorus trichloride (PCl_3) is formed. Phosphorus is a Group VA element, and chlorine is in Group VIIA.

   (4) Show how hydrogen sulfide (H_2S) is formed. Hydrogen is a Group IA element, and sulfur is a Group VIA element.

   (5) Show how the compound carbon dioxide (CO_2) is formed. Carbon is a Group IVA element, and oxygen is a Group VIA element. Be careful that each atom has eight electrons through sharing. Consider the possibility of using double bonds if you don't have enough electrons for all single bonds.

(6) Draw the Lewis structure for SO_3. Sulfur and oxygen are both Group VIA elements.

3. Write out the Lewis structure for HNO_2, nitrous acid. You may use a periodic table and follow problems in the text as a guide.

4. Draw Lewis structures for the following:

   (1) NH_4Cl
   (2) PO_4^{3-}
   (3) H_2O
   (4) MnO_4^-
   (5) CO_3^{2-}
   (6) ClO^-

5. If we examine many different compounds for the kinds of chemical bonds that hold them together, we will generally find two types. The two types of chemical bonds are called (1) _____ and the (2) _____ bond. When a transfer of electrons takes place from one atom to another, an (3) _____ bond is formed.

   A cation and an anion will form an (4) _____ bond since oppositely charged particles (5) attract/repel each other. Metallic elements tend to form ionic bonds when combining with the nonmetals, as we saw in the previous section.

   The predominant type of chemical bond is the (6) _____ bond. This type of bond occurs in the hydrogen molecule and develops as a result of each hydrogen atom contributing (7) one/two electron(s) to form the bond. The 1s electron orbitals of the hydrogen atoms overlap and pair to form a stable hydrogen molecule. There is a strong tendency for the hydrogen molecule to form from two individual atoms since in the molecule each (8) _____ charged electron is attracted to two (9) _____ charged nuclei.

The covalent bond is usually indicated by a dash mark (—). A single dash means (10) _____ pair of electrons and a double dash means (11) _____ pairs of electrons.

6. Given the following table of electronegativity values, indicate which of the listed binary compounds has polar covalent bonds. Also, calculate the difference in electronegativity values for one of the covalent bonds in each molecule.

Electronegativity Values

H	2.1	Br	2.8	Se	2.4
B	2.0	Cl	3.0	Te	2.1
P	2.1	F	4.0	N	3.0
O	3.5	S	2.5	Na	0.9

For example, the compound HF has one covalent bond. Is the bond between H and F polar? Yes, there is a significant difference in electronegativity values of $4.0 - 2.1 = 1.9$. Examine the remaining compounds in the same manner.

	Polar Covalent Bond (yes or no)	Electronegativity Value Difference
(1) H_2O	_____	_____
(2) PH_3	_____	_____
(3) BrCl	_____	_____
(4) H_2S	_____	_____
(5) OF_2	_____	_____
(6) NH_3	_____	_____
(7) H_2Se	_____	_____
(8) H_2Te	_____	_____
(9) Na_3P	_____	_____
(10) NCl_3	_____	_____
(11) Br_2	_____	_____

7. In the following, which of each pair will be larger?

(1) Cl^- and Cl^0 (3) Na^+ and Na^0
(2) Al^0 and Al^{3+} (4) Fe^{2+} and Fe^{3+}

8. Draw Lewis structures for the following:

   (1) H₂CO   (2) S₂O₃²⁻

9. Predict the type of bond that would be formed between the following pairs of atoms. Use Table 11.5 in the text.

   (1) Li & I   (2) C & H   (3) Al & N
   (4) Cs & P   (5) Se & S   (6) Ba & C

RECAP SECTION

Chapter 11 presents you with a great deal of useful information about the formation of chemical compounds from individual atoms. You now know how to describe chemical bonding in terms of electron transfer or electron sharing. And by using the concept of electronegativity, you can be more precise about the kind of electron sharing that takes place in covalent bonding. The concept of electronegativity is a most valuable tool in chemistry. The general theme of the unit has been to examine the formation of compounds and to describe the forces of attraction in compounds. The last sections in the chapter discuss the three-dimensional shapes of molecules and how to use a model called "valence shell electron pair repulsion" to predict molecular shape from Lewis structures.

ANSWERS TO QUESTIONS AND SOLUTIONS TO PROBLEMS

1. (1) positively   (2) ionization   (3) low   (4) high
    (5) high   (6) higher   (7) farther from (8) loses
    (9) cation   (10) 18   (11) anion   (12) gains
    (13) gaining   (14) eight   (15) losing   (16) gaining

- 91 -

2. (1) Zinc is a Group IIB element, which means that it has two valence electrons in its outer shell. Iodine is a Group VIIA element, which means that it has seven valence electrons in its outer shell. Zinc loses one electron to each of the iodine atoms and becomes a +2 charged cation. Each iodide ion thus has a −1 charge.

$$Zn + \begin{matrix} :\ddot{\underset{..}{I}}: \\ \cdot\ddot{\underset{..}{I}}: \end{matrix} \longrightarrow Zn^{2+} \quad \begin{matrix} :\ddot{\underset{..}{I}}:^- \\ :\ddot{\underset{..}{I}}:^- \end{matrix}$$

(2) $$\begin{matrix} K \cdot \\ +\ddot{O}: \\ K \cdot \end{matrix} \longrightarrow \begin{matrix} K^+ \\ :\ddot{\underset{..}{O}}:^{2-} \\ K^+ \end{matrix}$$

(3) $:\ddot{\underset{..}{Cl}}\cdot \longrightarrow \cdot\dot{P}\cdot \longleftarrow \cdot\ddot{\underset{..}{Cl}}: \longrightarrow :\ddot{\underset{..}{Cl}}:\ddot{P}:\ddot{\underset{..}{Cl}}:$
    $\uparrow \qquad\qquad :\ddot{\underset{..}{Cl}}:$
    $:\ddot{\underset{..}{Cl}}:$

(4) $H\cdot \longrightarrow \cdot\ddot{S}: \longrightarrow H:\ddot{\underset{..}{S}}:$
    $\quad\uparrow \qquad\qquad H$
    $\quad H$

(5) In order to write a Lewis structure for the molecule CO_2, which will have eight electrons around each atom, we must use double bonds between C and each oxygen atom. Each double bond consists of four electrons — two from the C atom and two from each oxygen atom.

$$:\ddot{O} \longrightarrow \cdot C: \quad \begin{matrix} \nearrow \ddot{O}: \\ \end{matrix} \longrightarrow :\ddot{O}::C::\ddot{O}:$$

The problem involves the formation of double bonds. Since one pair of electrons is one bond, there are two bonds between the two oxygen atoms and the carbon atom. Only in this manner will each atom have eight electrons in its outer shell.

(6) $$\begin{matrix} \nearrow:\ddot{O} \\ S + \\ \swarrow\searrow \\ :\ddot{O}:\ddot{O}: \end{matrix} \longrightarrow \begin{matrix} :\ddot{O}: \\ \| \\ S \\ \swarrow\searrow \\ :\ddot{O}:\ddot{O}: \end{matrix}$$

The Lewis structure for SO_3 involves one double bond and two single bonds in order to place eight electrons around each atom.

3. H:Ö:N̈::Ö: or H—O—N=O

   all covalent bonds   a "dash" equals a pair of electrons

Working out a Lewis structure for a somewhat complicated compound such as HNO_3 is a little bit like working a puzzle. We have four pieces to fit together so that each piece has a complete "set" of electrons — either two (as for H) or eight (as for O and N). Let's start by listing the number of electrons available.

N has five electrons, each O has six electrons, and H has one electron.

Take nitrogen first. We need three electrons to fill out the eight. It's common for oxygen to share one or two electrons with another atom.

If one O shares two electrons with nitrogen, then the other oxygen can share one with hydrogen and another with nitrogen. Using symbols, we have

H:Ö::N:Ö:

We should simplify this to show only two electrons on each of the four sides of the oxygen atoms and the nitrogen atom.

H:Ö:N̈:Ö:

With this situation the H, one O, and the N are happy with their number of electrons but the second O has only six. We have another possibility, which is to move one of the nonbonded pairs of electrons from N to go between the N and the O.

H:Ö:N̈::Ö:

Now all four atoms have filled orbitals and the bonds are covalent bonds. Another possibility you might have tried would be to bond the H to the nitrogen.

         H
:Ö::N::Ö:

This structure satisfies the requirement for the number of electrons around each atom but involves two double bonds, one each between the nitrogen and each oxygen. However, other evidence suggests that this is not the correct structure. Also, from other evidence we know that HNO_2 is an acid, which suggests the H is bonded to an oxygen atom.

4. (1) $\left[\begin{array}{c} H \\ H:N:H \\ H \end{array} \right]^+$ (2) $[:\ddot{Cl}:]^-$ $\left[:\ddot{O}:\overset{:\ddot{O}:}{\underset{:\ddot{O}:}{P}}:\ddot{O}: \right]^{3-}$

   (3) H:Ö:H   (4) $\left[:\ddot{O}:\overset{:\ddot{O}:}{\underset{:\ddot{O}:}{Mn}}:\ddot{O}: \right]^-$

(5)
$$\left[\begin{array}{c} :\ddot{O}: \\ \vdots\vdots \\ C \\ :\ddot{O}: \quad :\ddot{O}: \end{array}\right]^{2-}$$

There are 4e⁻ from the carbon, 6e⁻ each from the oxygen and 2 additional (negative charge on the ion is −2). So we have to distribute 24e⁻ between the central carbon and the 3 oxygen atoms. This means that one of the bonds needs to be shown as a double bond.

(6) $[:\ddot{C}l:\ddot{O}:]^{-}$

5. (1) ionic   (2) covalent   (3) ionic
    (4) ionic   (5) attract   (6) covalent
    (7) one   (8) negatively   (9) positively
    (10) one   (11) two

6.

Compound	Polar Covalent Bond (yes or no)	Electronegativity Value Difference
(1) H_2O	yes	1.4
(2) PH_3	no	0
(3) $BrCl$	yes	0.2
(4) H_2S	yes	0.4
(5) OF_2	yes	0.5
(6) NH_3	yes	0.9
(7) H_2Se	yes	0.3
(8) H_2Te	no	0
(9) Na_3P	yes	1.2
(10) NCl_3	no	0
(11) Br_2	no	0

7. (1) Cl^- will be larger   (3) No^0 will be larger
    (2) Al^0 will be larger   (4) Fe^{2+} will be larger

8.
    (1) $H:\overset{\overset{H}{\cdot\cdot}}{C}::\ddot{O}:$ (2) $[:\ddot{O}:\ddot{S}:\ddot{O}:\ddot{S}::\ddot{O}:]^{2-}$

9. (1) polar covalent   (2) polar covalent   (3) polar covalent
    (4) polar covalent   (5) nonpolar covalent   (6) ionic

WORD SEARCH 1

In the given matrix of letters, find the terms that match the following definitions. The terms may be horizontal, vertical, or on the diagonal. They may also be written forward or backward. Answers are found at the end of Chapter 20.

1. A subatomic particle with a charge of +1.
2. The mass of an object divided by its volume.
3. The basic building block of matter that cannot be broken down into simpler substances by ordinary chemical changes.
4. An electrically charged atom or group of atoms.
5. The central part of an atom
6. An element that is ductile and malleable.
7. The abbreviation for the name of an element.
8. The chemical law concerning the occurrence of the chemical properties of elements.
9. State of matter that is least compact.
10. Atoms of an element having the same atomic number but different atomic masses.
11. A subatomic particle with a charge of –1.
12. A negatively charged particle.
13. A small, uncharged individual unit of a compound.
14. Matter having uniform properties throughout.
15. The smallest particle of an element that can enter into a chemical reaction.
16. The relative attraction that an atom has for the electrons in a covalent bond.
17. A molecule with a separation of charge.
18. A subatomic particle that is electrically neutral.
19. The energy required to remove an electron from an atom.
20. A substance composed of two or more elements combined in a definite proportion by mass.
21. A solid without definite crystalline form.
22. A cloud-like region around the nucleus where electrons are located.
23. The metallic elements characterized by increasing numbers of d and f electrons in an inner shell.
24. Metric unit of length.

H	C	O	M	P	O	U	N	D	D	I	S	O	T	O	P	E	S
O	X	C	S	Z	K	L	O	Y	P	M	N	Q	R	B	L	C	T
M	P	R	U	D	K	Q	T	V	F	V	G	M	W	E	Z	T	Y
O	I	Z	E	O	R	V	O	M	T	S	E	O	C	U	A	L	G
G	U	A	L	T	E	I	R	O	R	E	S	T	T	L	A	C	I
E	I	F	C	S	U	P	P	W	L	L	R	A	B	A	I	T	U
N	G	O	U	Q	E	G	Y	O	H	O	J	Y	G	R	N	O	P
E	E	E	N	E	U	T	R	O	N	P	A	M	V	E	E	S	E
O	L	B	U	I	S	W	H	E	J	I	M	N	M	T	K	O	R
U	S	O	V	T	Z	U	G	J	F	D	V	E	C	L	D	E	I
S	I	E	I	R	K	A	O	A	B	K	L	T	U	M	L	E	O
K	N	G	S	D	T	M	T	H	V	E	P	R	S	D	D	A	D
L	L	A	T	I	B	R	O	I	P	A	D	O	I	M	E	R	I
O	E	W	V	S	L	H	F	L	O	R	G	E	U	Q	N	V	C
B	O	I	O	N	N	A	P	A	E	N	O	L	T	W	S	Z	L
M	T	V	M	E	O	S	I	D	N	C	E	M	N	O	I	N	A
Y	L	S	A	T	R	T	Z	M	A	B	U	N	A	L	T	U	W
S	P	J	D	A	T	C	I	O	T	M	J	L	E	K	Y	V	Z
A	E	O	R	B	C	M	R	P	E	B	U	M	E	R	E	I	D
B	H	T	L	C	E	A	H	T	M	X	Y	U	T	T	G	G	C
G	D	I	U	F	L	Q	A	O	F	E	S	N	X	G	A	Y	V
S	T	N	E	M	E	L	E	N	O	I	T	I	S	N	A	R	T
A	M	G	W	U	P	Q	N	S	W	D	L	E	R	V	K	C	B
H	I	V	R	E	T	E	M	O	R	D	Y	H	R	W	E	J	F

CHAPTER TWELVE

The Gaseous State of Matter

SELF-EVALUATION SECTION

1. The science of chemistry on a quantitative basis began with the systematic study of gas behavior, which may seem somewhat ironic since substances in the gas state are difficult to handle and use for experimental purposes. The fundamental properties of gases were established as compressibility, diffusion, pressure, and expansion. In addition, equal volumes of gases at the same temperature were found to exert identical pressures.

   The assumptions of the kinetic-molecular theory for an ideal gas are listed below.

   a. Gases consist of tiny particles.
   b. The volume of gas is mostly empty space.
   c. Gas molecules have no attraction for each other.
   d. Gas molecules move in straight lines in all directions, undergoing frequent collisions.
   e. No energy is lost through the collisions of gas molecules.
   f. The average kinetic energy for molecules is the same for all gases at the same temperature.

   Properties — list the letter for one or more assumptions from the above list that describe each of the following properties of gases.

   (1) Compressibility   _____
   (2) Diffusion   _____
   (3) Pressure   _____
   (4) Expansion   _____
   (5) Pressure of equal volumes of gases _____

2. An important assumption of the kinetic-molecular theory is that gas molecules move in straight lines and collide with each other and the walls of the container. Pressure, whether it's water pressure or gas pressure, is the force exerted on a unit area by a substance. As gas molecules collide with the container walls, they exert a certain pressure. What will happen to the pressure of a gas sample if more gas molecules are placed in a container? More molecules mean more collisions with the walls. Therefore, the pressure will rise. If the temperature and volume of the container are kept constant, there is a direct relationship

between pressure and number of gas molecules. Thus, doubling the amount of gas will double the pressure. Reducing the amount of gas to one-third the original quantity will reduce the pressure correspondingly. Refer to Figure 12.4 in the text.

3. Boyle's law and Charles' law.

What happens to the volume of the gas when the pressure is increased? Is the volume increased or reduced? We know from experience that gases can be compressed so that an increase in pressure will reduce the volume, but is there a quantitative or mathematical relationship that might be useful? Robert Boyle showed that, at constant temperature (T), the volume (V) of a gas is inversely proportional to the pressure (P). Thus, increased pressure reduced the volume, and reduced pressure results in increased volume. There are different ways to present Boyle's law, and you will be asked to choose the correct ways in a moment. Before we do that, what happens to a volume of gas when the absolute temperature increases? We know gases have kinetic energy. Since the mass of the gas molecules is constant, the velocity must increase. The question now is: Will the gas expand to a new volume? Yes, it will if the pressure of a gas can be maintained at a constant value. Thus, at constant pressure the volume of a gas is directly proportional to the absolute temperature. If the temperature goes up, the volume goes up, and vice versa. What happens to a gas sample if the pressure is not constant and the volume cannot expand? Visualize a closed empty can placed on a fire. The temperature of the gas increases; kinetic energy and pressure increase. The gas cannot expand so the pressure increases to the point at which the mechanical strength of the can cannot withstand the high pressure, and the can explodes.

Given below are several ways of mathematically stating Boyle's law and Charles' law. Place a "B" (for Boyle's) or "C" (for Charles') by each formula.

(1) $PV = k$

(2) $\frac{V_1}{T_1} = \frac{V_2}{T_2}$

(3) $V \propto \frac{1}{P}$

(4) $P_1 V_1 = P_2 V_2$

(5) $V \propto T$

(6) $\frac{V}{T} = k$

The symbol "$\propto$" means "to vary" or "to be proportional to".

4. Which of the gases in each pair will effuse at the fastest rate according to Graham's law?

(1) N_2, F_2

(2) CO_2, Ne

(3) SO_2, Br_2

(4) H_2, He

(5) Cl_2, O_2

(6) NH_3, CO

5. **Gas law problems**

   (1) The pressure of a gas in a piston is 2100 torr. The pressure is reduced to 550 torr. The original volume was 1.00 liter. What is the new volume?

   Do your calculations here.

   (2) A certain gas exerted a pressure of 785 torr in a 3.00 liter container. The gas was compressed into a 250. mL gas bottle. What was the new pres-sure?

   Do your calculations here.

   (3) 100. mL of a gas are obtained from a reaction at 300.°C. What volume will the gas occupy at 100.°C?

   Do your calculations here.

(4) A 250. mL sample of gas originally at −50.°C was brought to room temperature (25°C). Find the new volume.

Do your calculations here.

(5) You should also be able to work a combined gas law problem that involves P, V, and T. Try this one.

What is the volume of a gas at STP if the original conditions were 45 mL, 0.75 atm pressure, and 21°C?

Do your calculations here.

6. Fill in blank or circle the appropriate response.

We have learned in Chapter 13 that at standard temperature and pressure, which are (1) _____ and (2) _____ , 1 mole of any gas occupies (3) _____ liters. This volume is known as the molar volume and is constant for any gas at STP. Another way to state the relationship is to say that Avogadro's number of gas molecules (6.022×10^{23}) at STP is equal to 1 molar volume. When performing calculations involving gases, the usual practice is to convert all volumes of gases to standard conditions. Then the relationship 1 mole = 22.4 liters can be used for stoichiometric calculations. As a summary statement about gas behavior, one might state Avogadro's Law, which says that equal volumes of gases at the (4) same/different temperature and pressure contain equal numbers of molecules. This statement follows from the assumption that two gases at the same temperature possess the same average kinetic energy. If they occupy the same volume, the pressures will be the same. This is a consequence of the same number of molecules moving with the same average kinetic energy.

7. The density of a gas is expressed in grams per liter.

$$d = \frac{\text{mass}}{\text{volume}} = \frac{g}{L}$$

At STP, density may also be calculated from the following equation.

$$\text{density at STP} = \frac{\text{molar mass}}{22.4 \text{ L/mol}} = \frac{g/\text{mol}}{\text{liters/mol}} = \frac{g}{L}$$

(1) At STP, what is the density of NO (nitrogen oxide) gas? Use your table of atomic masses.

Do your calculations here.

(2) The density of HCN (hydrogen cyanide) gas at STP is 1.21 g/L. What volume will 100. g of HCN occupy at STP?

Do your calculations here.

(3) A quick way to prepare acetylene gas in the laboratory is to drop chunks of calcium carbide into water. Old-fashioned miners' lamps used this reaction, as did some residential lighting systems years ago. We will work a Dalton's partial pressure problem from this reaction.

A sample of C_2H_2 (acetylene) gas collected over water at 21°C and 750. torr pressure occupies a volume of 175 mL. Calculate the volume of dry acetylene at STP. The vapor pressure of water at various temperatures is listed in Appendix II of the textbook. Dalton's law states that, in a mixture of gases, each gas exerts its own individual pressure.

$$P_{total} = P_A + P_B + P_C$$

For this particular problem, the total pressure of the moist acetylene collected is made up of two parts — the pressure of C_2H_2 and the pressure of the water vapor. We have to subtract the partial pressure of the water vapor from the C_2H_2 before we correct the C_2H_2 volume to STP.

Do your calculations here.

8. What volume of NO gas will be produced from 4.00 moles of N_2 and 3.00 moles of O_2 at STP accordign to the following equation?

$$N_2(g) + O_2(g) \longrightarrow 2\ NO(g)$$

Do your calculations here.

9. What volume of NO at STP gas can be produced from 7.0 moles of nitrogen dioxide (NO_2) according to the following equation?

$$3\ NO_2 + H_2O \longrightarrow 2\ HNO_3 + NO$$

Do your calculations here.

10. Using the ideal gas equation, calculate the volume that 15 g of H_2 gas at 25°C and 1.2 atm pressure will occupy.

$$PV = nRT$$

Do your calculations here.

Challenge Problems

11. In a movie thriller, the hero is locked in a sealed room. The bad guys allow a mixture of two poisonous gases, arsine (AsH_3 – garlic odor) and cyanogen (C_2N_2 – almond odor), to effuse through a porous opening into the room. If the hero doesn't find an escape route, what will be the fragrance that reaches the nostrils first?

Do your calculations here.

12. The size cyclinder known as a 1A cylinder has a volume 43.8 L. How many moles of oxygen are contained in a 1A cylinder at 72°F and 1500 pounds per in^2 pressure (psi)? To convert psi to atmospheres multiply psi by 0.06805 atm/psi. What is the mass of this number of moles of oxygen gas?

Do your calculations here.

13. Use the ideal gas equation to calculate the molar mass of arsine (the garlic odor from problem 11), given the following information.

> 3.48 g of AsH_3 at 740. torr and 21°C occupies 1.11 L of volume. What is the molar mass of AsH_3?

Do your calculations here.

RECAP SECTION

Chapter 12 is long and contains a great deal of new material. However, you should have sharpened up your problem-solving skills and learned a great deal about the gaseous state. Particularly important is the ability to visualize in equation form a word statement such as Boyle's law and then use the relationship to solve problems. The relationships of observed gas properties to the kinetic-molecular theory of ideal gas behavior, Avogadro's Law, Gay-Lussac's law of combining volumes, and Dalton's Law of Partial Pressures were studied. However, it is important to realize that real gases don't exactly behave as the ideal gas equation predicts because real gases do have finite volume and also exhibit intermolecular attractions, particulary at low temperatures and high pressures. You should feel confident in tackling almost any stoichiometric problem, given an equation and an atomic mass table. You will use these skills again in later chapters in solution chemistry.

ANSWERS TO QUESTIONS AND SOLUTIONS TO PROBLEMS

1. (1) Compressibility – b
   (2) Diffusion – d
   (3) Pressure – d
   (4) Expansion – b, c, d
   (5) Pressure of equal volumes of gases – f, e

2. No answer required

3. (1) B   (2) C   (3) B   (4) B   (5) C   (6) C

4. (1) N_2 (2) Ne   (3) SO_2 (4) H_2 (5) O_2 (6) NH_3

5. (1) 3.8 liters

The problem involves pressures and volumes; this means a Boyle's law problem.

The original volume was 1.00 L, and the original pressure was 2100 torr. The pressure dropped to 550 torr. What happens to the volume of a gas when the pressure goes down? According to Boyle's law, the volume increases. This means we must multiply the original volume by a ratio of pressures, which will give us a larger volume than 1.00 L.

original volume × ratio of pressure = new volume

$$1.00 \text{ L} \times \frac{2100 \text{ torr}}{550 \text{ torr}} = 3.8 \text{ L}$$

(2) 9.42×10^3 torr

This time let's approach the problem using Method (b) (algebraic) from the chapter. Organizing the data and converting the units to be the same gives:

$P_1 = 785$ torr $\qquad P_2 = ?$

$V_1 = 3.00$ L $\qquad V_2 = 250$ mL $= 0.250$ L

Using Boyle's Law:

$$P_1 V_1 = P_2 V_2$$

and dividing both sides of the equation by V_2:

$$\frac{P_1 V_1}{V_2} = P_2$$

Now substituting the values from the data table:

$$\frac{(785 \text{ torr})(3.00 \text{ L})}{0.250 \text{ L}} = 9.42 \times 10^3 \text{ torr}$$

(3) 65.1 mL

The problem involves volumes and temperatures, which makes it a Charles' law problem. We know that V and T are directly related as long as T is expressed as a Kelvin or absolute temperature. We first have to convert °C to K. Any time you are asked to solve a gas law problem with a temperature involved, be sure to convert to K.

300.°C + 273.15 = 573 K

100.°C + 273.15 = 373 K

The temperature dropped from 573 K to 373 K. This means the volume must decrease also.

original volume × ratio of temperatures = new volume

$$100. \text{ mL} \times \frac{373 \text{ K}}{573 \text{ K}} = 65.1 \text{ mL}$$

(4) 334 mL

Once again we will use Method (b) (algebraic) from the chapter. Organizing the data and converting the temperatures to Kelvin gives:

$V_1 = 250.\ \text{mL}$ $\qquad V_2 = ?$

$T_1 = -50.°C = 223\ \text{K}$ $\qquad T_2 = 25°C = 298\ \text{K}$

This is Charles' Law problem so,

$$\frac{V_1}{T_1} = \frac{V_2}{T_2}$$

Multiplying both sides of the equation by T_2 gives $\frac{V_1 T_2}{T_1} = V_2$. Substituting

$$\frac{(250.\ \text{mL})(298\ \cancel{K})}{223\ \cancel{K}} = 334\ \text{mL}$$

(5) With a problem like this, it might be well to tabulate the data in an organized fashion.

$P_1 = 0.75\ \text{atm}$ $\qquad P_2 = 1.0\ \text{atm (standard)}$

$V_1 = 45\ \text{mL}$ $\qquad V_2 = ?$

$T_1 = 21°C = 294\ \text{K}$ $\qquad T_2 = 0°C = 273\ \text{K (standard)}$

Since P, V, and T are all changing, the combined gas law is used

$$\frac{P_1 V_1}{T_1} = \frac{P_2 V_2}{T_2}$$

Solving of V_2 and substituting data:

$$\frac{P_1 V_1 T_2}{T_1} = V_2 = \frac{(0.75\ \cancel{\text{atm}})(45\ \text{mL})(273\ \cancel{K})}{(1.0\ \cancel{\text{atm}})(294\ \cancel{K})} = 31\ \text{mL}$$

6. (1) 0°C, 273 K (2) 1 atm, 760 torr (3) 22.4 L (4) same

7. (1) 1.34 $\frac{g}{L}$

The molar mass of NO is 14.01 g/mol + 16.00 g/mol = 30.01 $\frac{g}{\text{mol}}$

$$d = \frac{30.01\ \text{g}}{\cancel{\text{mol}}} \times \frac{1\ \cancel{\text{mol}}}{22.4\ \text{L}} = 1.34\ \frac{g}{L}$$

(2) 82.6 L

Solve the equation for volume (V).

$$d = \frac{m}{V}$$

Multiply each side of the equation by "V".

$$(V)(d) = \frac{(m)(V)}{(V)}$$

Now divide each side by d

$$\frac{(V)(d)}{(d)} = \frac{m}{d}$$

We can now substitute the values given for the density and the mass into the equation.

$$V = \frac{100. \text{ g} \times 1 \text{ L}}{1.21 \text{ g}} = 82.6 \text{ L HCN}$$

(3) 156 mL

Since the acetylene gas was collected over water, we have to subtract the partial pressure of water vapor at 21°C to find the partial pressure of acetylene alone. Then, we can find the volume at STP.

$$P_{total} = P_{C_2H_2} + P_{H_2O}$$
$$750. \text{ torr} = P_{C_2H_2} + 18.6 \text{ torr (Appendix II)}$$
$$P_{C_2H_2} = 750. \text{ torr} - 18.6 \text{ torr} = 731 \text{ torr}$$

Establish a table

$P_1 = 731.4$ torr $P_2 = 760.$ torr (standard)
$V_1 = 175$ mL $V_2 = x$ mL
$T_1 = 21°C = 294$ K $T_2 = 0°C = 273$ K (standard)

Using the ratio technique, we see that pressure is increased and the temperature is reduced, both of which reduce the volume.

$$175 \text{ mL} \times \frac{731 \text{ torr}}{760. \text{ torr}} \times \frac{273 \text{ K}}{294 \text{ K}} = 156 \text{ mL}$$

8. 134 L

This is a volume-volume calculation and is an application of Avogadro's Law. Also, we need to consider whether there is a limiting reactant. First, determine how many moles of NO$_{(g)}$ can be produced from each reactant.

$$4.00 \text{ mol } N_2 \times \frac{2 \text{ mol NO}}{1 \text{ mol } N_2} = 8.00 \text{ mol NO}$$

$$3.00 \text{ mol } O_2 \times \frac{2 \text{ mol NO}}{1 \text{ mol } O_2} = 6.00 \text{ mol NO}$$

Therefore, O_2 is the limiting reactant and 6.00 mol NO will be produced in the reaction with 1.00 mol of N_2 left over.

The volume of 6.00 mol of NO at STP will be

$$6.00 \text{ mol} \times 22.4 \frac{\text{L}}{\text{mol}} = 134 \text{ L}$$

9. 52 L

   The number of moles of starting substance is 7.0 moles NO_2.

   Calculate the moles of NO, using the mole-ratio method.

   $$7.0 \text{ mol } NO_2 \times \frac{1 \text{ mol NO}}{3 \text{ mol } NO_2} = 2.3 \text{ mol NO}$$

   Convert moles of NO to L of NO. The moles of a gas at STP are converted to L by multiplying by the molar volume, 22.4 L per mole:

   $$2.3 \text{ mol NO} \times \frac{22.4 \text{ L}}{\text{mol}} = 52 \text{ L NO}$$

10. 1.5×10^2 L (2 significant figures)

    We must first calculate how many moles 15 g of H_2 represents and then solve the equation for "volume".

    $$15 \text{ g} \times \frac{1 \text{ mol } H_2}{2.0 \text{ g}} = 7.5 \text{ mol } H_2$$

    $$PV = nRT \quad \text{or} \quad V = \frac{nRT}{P}$$

    $$R = 0.0821 \text{ L} \cdot \text{atm/mol} \cdot \text{K}$$

    $$V = \frac{7.5 \text{ mol} \times 0.0821 \frac{\text{L} \cdot \text{atm}}{\text{mol} \cdot \text{K}} \times 298 \text{ K}}{1.2 \text{ atm}}$$

    $$= 1.5 \times 10^2 \text{ L}$$

11. Almond odor

    The molar mass of arsine is 77.94 g/mol and that of cyanogen is 52.04 g/mol. since the rate of effusion is inversely proportional to the square roots of their molar masses, the hero will smell almonds first.

12. 1.80×10^2 moles O_2 and 5.76×10^3 g O_2

    We need to use the ideal gas equation to solve the problem. The data tabulated looks like this:

    $$P = (1500 \text{ psi}) = (1500 \text{ psi})\left(0.06805 \frac{\text{atm}}{\text{psi}}\right) = 1.0 \times 10^2 \text{ atm}$$

    $$V = 43.8 \text{ L}$$

    $$R = 0.0821 \frac{\text{L} \cdot \text{atm}}{\text{mol} \cdot \text{K}}$$

    $$T = 72°F = 22°C = 295 \text{ K}$$

    $$PV = nRT$$

    $$n = \frac{PV}{RT} = \frac{(1.0 \times 10^2 \text{ atm})(43.8 \text{ L})}{\left(0.0821 \frac{\text{L} \cdot \text{atm}}{\text{mol} \cdot \text{K}}\right)(295 \text{ K})} = 1.80 \times 10^2 \text{ mol } O_2$$

The mass of oxygen equals

$$(1.80 \times 10^2 \text{ mol } O_2)(32.00 \text{ g/mol}) = 5.76 \times 10^3 \text{ g}$$

13. 77.7 g/mol

    First, we need to convert torr into atm and °C into K.

    $$740. \text{ torr} \times \frac{1 \text{ atm}}{760. \text{ torr}} = 0.974 \text{ atm}$$

    $$21°C + 273.15 = 294 \text{ K}$$

    The ideal gas equation is: $PV = nRT$

    Rearranging and using grams/molar mass = n, we have

    $$\text{molar mass (M)} = \frac{g}{PV} RT$$

    Substituting into the equation:

    $$M = \frac{3.48 \text{ g} \times 0.0821 \text{ L·atm} \times 294 \text{ K}}{0.974 \text{ atm} \times 1.11 \text{ L} \times \text{mol·K}}$$

    $$= 77.7 \text{ g/mol}$$

    This value varies somewhat from the accepted value of 77.94 g/mol because of rounding off several of the given values.

CHAPTER THIRTEEN

Water and the Properties of Liquids

SELF-EVALUATION SECTION

1. Many of the important characteristics of water are related to its physical properties. For example, the amount of heat required to change 1 gram of a solid into a liquid, called the (1) _____ , is an unusually large amount for water, as is the amount of heat required to change 1 gram of liquid at its normal boiling point into a gas called the (2) _____ . The temperature at which ice begins to change into the liquid state is called the (3) _____ and the temperature at which the vapor pressure of water equals the atmospheric pressure of 1 atm is called the (4) _____ .

Water reacts with numerous compounds to form useful products. For example, certain metallic oxides react with water to form bases and are known as (5) _____ , while nonmetallic oxides form acids with water and are known as (6) _____ . Calcium oxide (CaO) is a basic anhydride which is used in the cement industry, and sulfur trioxide (SO_3) is the acid anhydride of sulfuric acid (H_2SO_4). Other compounds contain water molecules as part of their crystalline structure and are called (7) _____ . One often sees dramatic color changes when the water of hydration is removed from a hydrated salt. Some compounds are capable of absorbing water directly from the atmosphere. These substances are said to be (8) _____ . Other substances, called (9) _____ , absorb enough water to form a solution. Sodium hydroxide and P_2O_5 are examples of deliquescent substances.

2. Water, which is the most common chemical substance around us, has been the subject of many experiments over the years. It is a simple molecule, H_2O, and yet its properties suggest that water is a very large molecule. We want to analyze how the structure of water influences such properties as boiling point and melting point. Examine Table 13.3 in the text prior to answering this section.

Fill in the blank space or circle the appropriate response.

A single water molecule consists of (1) _____ H atom(s) and (2) _____ O atom(s). The oxygen atom is the middle atom joined to each H atom by a(n) (3) ionic/covalent bond. The molecule is (4) straight end to end/bent in the middle with a bond angle of 105° between the two covalent bonds.

Oxygen is a very (5) electropositive/electronegative element and, as a result, the two covalent OH bonds are (6) nonpolar/polar. The bend in the molecule in conjunction with the polar covalent bonds make water a (7) nonpolar/polar molecule. The oxygen atom carries a partial (8) _____ charge, and each hydrogen atom carries a partial (9) _____ charge.

Since each water molecule has the same unequal charge distribution, there will be an attraction between molecules. The oxygen side of the molecule will be attracted to the (10) _____ atom of another water molecule through a weak electrostatic bond. This type of attraction between an H atom and a highly electronegative atom is called a (11) _____ bond. The weak ionic association between water molecules produces the effect of water behaving as a large molecule. When we examine the physical properties of water, it is clear that water does not fit the expected pattern. The melting point and normal boiling point are (12) higher/lower than expected, as are the heat of fusion and heat of vaporization. Water acts though it were a large bulky molecule rather than a small one with a molar mass of 18.02 g/mol. The bent structure and resulting polarity have a marked influence on the physical properties of water.

3. Identify each of the salt formulas as anhydrous or hydrates. Name them.
    (1) $CaSO_4$ _____
    (2) $CoCl_2 \cdot 6\ H_2O$ _____
    (3) $NaC_2H_3O_2 \cdot 3\ H_2O$ _____
    (4) K_2S _____
    (5) Na_3PO_4 _____

4. Identify the compounds listed as basic anhydrides or acidic anhydrides and write their reactions with water.

   (1) CaO
   (2) SO_3
   (3) N_2O_5
   (4) Na_2O

   Write your answers below.

5. Write the formulas for the anhydrides of the following.

   (1) H_2SO_3, $HClO_4$, H_2CO_3
   (2) KOH, $Ba(OH)_2$, $Mg(OH)_2$

   Write your answers below.

6. Complete and balance the following reactions involving water as a reactant.

   (1) $K(s) + H_2O(l) \rightarrow H_2(g) +$
   (2) $Al(s) + H_2O(\text{steam}) \rightarrow \qquad + Al_2O_3(s)$
   (3) $Fe(s) + H_2O(\text{steam}) \rightarrow H_2(g) +$
   (4) $Cl_2(g) + H_2O(l) \rightarrow \qquad \rightarrow HOCl(aq)$

7. Fill in the blank space.

Natural fresh waters are usually not pure enough to drink and therefore must be treated. Removal of large objects is accomplished by (1) _____ whereas fine particles are removed by (2) _____ and (3) _____ . The last step (4) _____ kills bacteria. If it contains dissolved magnesium and calcium salts, the water is said to be (5) _____ . Three techniques used to soften hard water are (6) _____ , (7) _____ , and (8) _____ . The process that uses zeolite to soften hard water is a type of (9) _____ technique.

8. Fill in the blank space or circle the appropriate response.

All substances in the liquid state are in the process of vaporizing. Some chemicals, such as acetone and ether, evaporate quickly, whereas others, such as mercury, do so slowly. The process of molecules going from the liquid state to the gas state is called (1) _____ . In any sample of a liquid, (2) repulsive/attractive forces exist that must be overcome before a molecule can escape the liquid state. Even though the temperature of the liquid is uniform, not all of the molecules possess the same (3) kinetic/potential energy. Since the masses are constant, this suggests that the velocities of the molecules are (4) the same/different. Therefore, molecules at the surface of the liquid, which are moving faster than their neighbors, are able to overcome the attractive forces and escape to the gas state. After these molecules with greater kinetic energy leave the liquid state, the average kinetic energy of the remaining molecules in the liquid state is (5) lowered/raised. Therefore, the temperature (6) raises/drops , and we find that vaporization is a (7) warming/cooling process, which we know to be true from everyday experience. If the temperature drops as the average kinetic energy of the system goes down, you might wonder why the process goes on until the liquid is gone.

In an open dish of water, for example, the water vaporizes completely. Where does the heat energy come from to maintain kinetic energy of the water so that vaporization proceeds? As you probably surmised, it comes from the surroundings.

What will happen to our open dish of water if we place a large glass cover over it? The molecules of water at the surface of the liquid with sufficient kinetic energy will enter the gas state, and the liquid level in the dish will do down very slightly until an equilibrium situation is reached. In this equilibrium state, just as many water molecules leave the liquid state as enter it from the saturated vapor inside the glass cover. If we increase the temperature and, therefore, the kinetic energy, more liquid will vaporize until a new equilibrium is established. If the temperature is reduced, water molecules in the vapor or gas state will condense to the liquid state and the water level will increase. At any temperature, once equilibrium is established, as many molecules leave the liquid state as enter from the gas state. When an equilibrium state is reached, the vapor exerts a pressure just as any other gas does.

This pressure is known as (8) _____ and is an "internal pressure" or measure of the escape tendency of molecules from the liquid to the gas state. When the vapor pressure of a liquid reaches atmospheric pressure (760 torr), the liquid begins to (9) _____. For water, the temperature at which this happens is (10) _____. The vapor pressure of ethyl ether reaches 760 torr between 30°C and 40°C; and ethyl alcohol reaches 760 torr between 70°C and 80°C. It should be noted that atmospheric pressure is not always 760 torr, but varies with the weather and the elevation above sea level.

9. On the following heating curve, a solid substance at point A is heated until it reaches point E. Identify the various stages along the curve as requested.

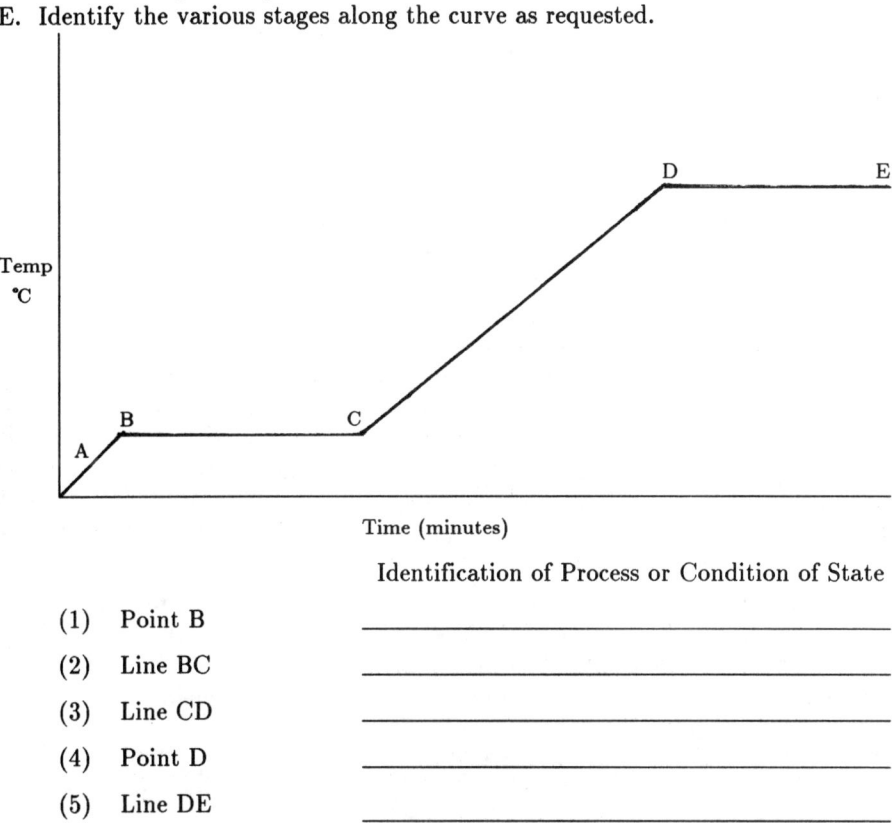

Identification of Process or Condition of State

(1) Point B   _____
(2) Line BC   _____
(3) Line CD   _____
(4) Point D   _____
(5) Line DE   _____

Challenge Problems

10. An ice cube at 0°C has a mass of 8.75 g. Calculate how much heat energy (joules) is needed to melt the ice cube, raise the temperature of the resulting water to 100°C, and convert it to steam at 100°C.

    The heat of fusion is 335 J/g. The heat of vaporization is 2.26 kJ/g.

    Do your calculations here.

11. How many calories of heat energy are required to vaporize two 25.0 g pieces of solid ethanol, C_2H_5OH, at $-112°C$ given the following information.

   Ethanol, C_2H_5OH melting point $-112°C$
   boiling point $78.4°C$
   Heat of fusion 24.9 cal/g
   Heat of vapoization 204.3 cal/g

   specific heat, C_p, $2.49 \frac{joule}{g \cdot °C}$

   $calories = (joules)\left(0.239 \frac{cal}{joule}\right)$

   Do your calculations here.

RECAP SECTION

Chapter 13 is an enjoyable chapter to read. Even though there are many new terms, everything seems to fit together. We discussed a chemical compound that we are in contact with every day. Much of the discussion centered around the application of previously learned concepts. The importance of electronegativity in explaining the physical properties of water was good exercise for you. Since some of the topics in the chapter were not covered in the study guide, you may want to go back over the sections on the chemistry of water on your own. Now that you have learned some things about water by itself, let us begin to use water as we do in the laboratory to prepare chemical solutions. In the next unit, we discuss the principal use of water in chemistry — as a solvent for reagents.

ANSWERS TO QUESTIONS

1. (1) heat of fusion   (2) heat of vaporization   (3) melting point
    (4) normal boiling point   (5) basic anhydrides   (6) acid anhydrides
    (7) hydrates   (8) hygroscopic   (9) deliquescent

2. (1) 2   (2) 1   (3) covalent
    (4) bent in the middle   (5) electronegative   (6) polar
    (7) polar   (8) negative   (9) positive
    (10) hydrogen   (11) hydrogen   (12) higher

3. (1) anhydrous calcium sulfate
   (2) hydrate cobalt (II) chloride hexahydrate
   (3) hydrate sodium acetate trihydrate
   (4) anhydrous potassium sulfide
   (5) anhydrous sodium phosphate

4. (1) CaO – basic anhydride
   $CaO + H_2O \rightarrow Ca(OH)_2$
   (2) SO_3 – acidic anhydride
   $SO_3 + H_2O \rightarrow H_2SO_4$
   (3) N_2O_5 – acidic anhydride
   $N_2O_5 + H_2O \rightarrow 2\ HNO_3$
   (4) Na_2O – basic anhydride
   $Na_2O + H_2O \rightarrow 2\ NaOH$

5. (1) SO_2 Cl_2O_7 CO_2
   (2) K_2O BaO   MgO

6. (1) $2\ K(s) + 2\ H_2O(l) \rightarrow H_2(g) + 2\ KOH(aq)$
   (2) $2\ Al(s) + 3\ H_2O(steam) \rightarrow 3\ H_2(g) + Al_2O_3(s)$
   (3) $3\ Fe(s) + 4\ H_2O(steam) \rightarrow 4\ H_2(g) + Fe_3O_4(s)$
   (4) $Cl_2(g) + H_2O(l) \rightarrow HCl(aq) + HOCl(aq)$

7. (1) screening (2) flocculation (3) sedimentation
   (4) disinfection (5) hard (6) precipitation
   (7) distillation (8) ion exchange (9) ion exchange

8. (1) vaporization (2) attractive (3) kinetic
   (4) different (5) lowered (6) drops
   (7) cooling (8) vapor pressure (9) boil
   (10) 100°C

9. (1) melting point
   (2) solid in equilibrium with liquid
   (3) all-liquid state, temperature begins to rise
   (4) boiling point
   (5) boiling liquid in equilibrium with gas

10. 26.4 kJ

    To melt the ice cube requires 335 J for each gram. Therefore,

    $\left(335\ \frac{J}{g}\right) \times 8.75\ g = 2930\ J$

    We now have water at 0°C and will need 4.184 J per gram of water per degree to raise the water temperature to boiling. Therefore,

$$(8.75 \cancel{g})(100.\cancel{°C})\left(4.184 \frac{J}{\cancel{g°C}}\right) = 3660.\ J$$

Notice how the units cancel to give us units of heat energy – joules. Now we must use 2.26 kJ for each gram of hot water to convert it into steam.

$$(8.75\ g)(2.26\ \tfrac{kJ}{g}) = 19.8\ kJ$$

The last step is to add up the individual values after converting to kJ.

$$2.93\ kJ + 3.66\ kJ + 19.8\ kJ = 26.4\ kJ$$

11. 1.71×10^4 calories

   The first step is to melt the two 25.0 g pieces of solid ethanol at –112°C.

   $$(50.0\ g)(24.9\ cal/g) = 1.25 \times 10^3\ cal\ (3\ significant\ figures)$$

   The temperature must be raised to the boiling point, 78.4°C. The specific heat is given in joules and must be changed to calories.

   $$(2.49\ joule/°C\ g)(0.239\ cal/joule)(50.0\ g)(temperature\ change)$$

   $$(2.49\ joule/°C\ g)(0.239\ cal/joule)(50.0\ g)(190.4°C) = 5.67 \times 10^3\ cal$$

   To vaporize the 50.0 g of liquid at the boiling point requires 204.3 cal/g

   $$(50.0\ g)(204.3\ cal/g) = 1.02 \times 10^4\ cal$$

   Adding up the 3 values gives the final answer

   $$(1240\ cal) + (5670\ cal) + (10{,}200\ cal) = 1.71 \times 10^4\ cal$$

CHAPTER FOURTEEN

Solutions

SELF-EVALUATION SECTION

1. In the following statements, identify the (a) solution, (b) solute, and (c) solvent.

   (1) Household bleach is a 5% solution of sodium hypochlorite (NaClO) in water.

   a. _____ b. _____ c. _____

   (2) A 0.1 M iodine solution was prepared by dissolving crystals of iodine in carbon tetrachloride (CCl_4).

   a. _____ b. _____ c. _____

   (3) Air is composed primarily of two gases — oxygen and nitrogen. Air is approximately 79% nitrogen and 21% oxygen.

   a. _____ b. _____ c. _____

   (4) Nickel coins in the United States are made from a nickel-copper alloy that is 75% copper and 25% nickel.

   a. _____ b. _____ c. _____

   (5) In order for certain species of fish to thrive in lakes and streams, the dissolved oxygen content of the water has to be 0.0005% or greater.

   a. _____ b. _____ c. _____

2. Solubility of salts in water.

   All nitrates are soluble in water.

   All chlorides, bromides, and iodides are soluble in water except those of silver, mercury (I), and lead (II).

   All carbohydrates and phosphates are insoluble in water except those of sodium, potassium, and ammonium.

   All sulfides are insoluble in water except those of ammonium sulfide and Group IA and Group IIA metals.

Place an *s* (for soluble) or *i* (for insoluble) next to each of these compounds.

(1) $CaBr_2$ _____ (11) $(NH_4)_2CO_3$ _____
(2) NH_4NO_3 _____ (12) BaI_2 _____
(3) Na_2S _____ (13) $LiI \cdot 3\ H_2O$ _____
(4) $Mg_3(PO_4)_2$ _____ (14) $PbCl_2$ _____
(5) AgI _____ (15) KCl _____
(6) Fe_2S_3 _____ (16) Na_2CO_3 _____
(7) $ZrCl_4$ _____ (17) $Sn(NO_3)_4$ _____
(8) $NiCO_3$ _____ (18) Ag_2S _____
(9) $Al(NO_3)_3 \cdot 9\ H_2O$ _____ (19) K_3PO_4 _____
(10) $HgBr$ _____ (20) $CaCO_3$ _____

3. Solubilities of salts in g/100 g H_2O.

$CaBr_2$	125 g at 0°C	$(NH_4)_2CO_3$	100 g at 15°C
NH_4NO_3	118 g at 0°C	$NaC_2H_3O_2$	119 g at 0°C
Na_2S	15.4 g at 10°C	$LiI \cdot 3\ H_2O$	151 g at 0°C
$Na_2S_2O_3$	50 g at 20°C	KCl	34.7 g at 20°C
$MgSO_4$	26 g at 0°C	Na_2CO_3	7.1 g at 0°C
$Al(NO_3)_2$	63.7 g at 25°C	K_3PO_4	90 g at 20°C

Identify the solutions as saturated, unsaturated, or supersaturated. (All solutions are in 100 g H_2O.)

(1) 5 g of $CaBr_2$ at 0°C
(2) 100 g of NH_4NO_3 at 0°C
(3) 15.4 g of Na_2S at 10°C
(4) 55 g of $Na_2S_2O_3$ at 20°C
(5) 4 g of $MgSO_4$ at 0°C
(6) 51 g of $Al(NO_3)_3$ at 25°C
(7) 75 g of $(NH_4)_2CO_3$ at 15°C
(8) 125 g of $NaC_2H_3O_2$ at 0°C
(9) 38 g of $LiI \cdot 3H_2O$ at 0°C
(10) 34.7 g of KCl at 20°C
(11) 6.5 g of Na_2CO_3 at 0°C
(12) 12 g of K_3PO_4 at 20°C

4. Fill in the blank space or circle the appropriate response.

Water molecules, which are very (1) <u>polar/nonpolar</u>, are (2) <u>attracted/repulsed</u> by other polar or ionic molecules or ions. Water molecules weaken the ionic forces that hold ions such as Na^+ and Cl^- together. Then the ions are pulled apart by the interaction with water as the water molecules surround the ions. The charged ions such as Na^+ and Cl^- then diffuse slowly away from the mass of undissolved salt as hydrated ions.

With most solid chemicals, an increase in temperature means an (3) _____ in solubility. However, a gaseous chemical always (4) _____ in solubility as the temperature increases. A temperature increase means that the kinetic energy of the gas molecules (5) _____ and their ability to associate with the solvent molecules (6) _____ .

Pressure changes don't affect the solubility of solids greatly but gases show marked changes. The solubility of a gas is (7) <u>inversely/directly</u> proportional to the pressure of the gas above the liquid. Double the pressure means the solubility will (8) _____ . Carbonated beverages are a good example.

5. List the four factors that influence the rate or speed at which a solid solute dissolves.

6. Before proceeding with exercises 6 through 10, be sure to study section 14.6 in the textbook.

   (1) How would you prepare 1500 g of a 2.000% sucrose (sugar) mass percent solution to be used for intravenous feeding in a hospital?

   Do your calculations here.

(2) What masses of sodium chloride (NaCl) and water are needed to make 525.0 g of 0.1000% salt solution?

Do your calculations here.

7. (1) Table wine is generally 12% alcohol by volume. How many mL of alcohol are contained in 250 mL of wine?

Do your calculations here.

(2) It is sometimes necessary in a chemistry laboratory to prepare a very corrosive solution of NaOH. How many grams of NaOH are contained in 375 g of 40.% NaOH solution?

Do your calculations here.

8. (1) What is the molarity of a solution containing 4.2 moles of H_2SO_4 in 300. mL of solution?

Do your calculations here.

(2) What is the molarity of a KCl solution containing 1.7 moles of KCl in 3.0 L of solution?

9. Calculate how many grams of chemical would be required to prepare the following solutions.

Atomic Masses, amu

(1) 600. mL of 0.15 M NaF
(2) 4.00 L of 8.00 M NH_4NO_3
(3) 420 mL of 0.70 M $Ca(OH)_2$
(4) 250 mL of 0.94 M Na_2SO_4
(5) 0.50 L of 3.00 M $NaNO_3$

Na = 22.99
S = 32.06
O = 16.00
N = 14.01

F = 19.00
Ca = 40.08
C = 12.01
H = 1.008

Use the relationship

$$M = \frac{\text{g of solute}}{\text{molar mass solute} \times \text{L of solution}}$$

or solve by dimensional analysis. **Do your calculations here.**

10. More experience in working with molarities of solutions.

   (1) What volume of 1.33 M KMnO$_4$ can be prepared from 180. g of KMnO$_4$?

   Atomic Masses, amu
   K = 39.10
   Mn = 54.94
   O = 16.00

   Do your calculations here.

(2) What volume of 0.10 M H_2SO_4 can be prepared from 147 g of H_2SO_4?

$$\text{Atomic Masses, amu}$$
$$H = 1.008$$
$$S = 32.06$$
$$O = 16.00$$

Do your calculations here.

(3) Calculate the number of moles of hydrochloric acid in 4.0 L of 0.333 M HCl.

Do your calculations here.

(4) Calculate the number of moles contained in the following solutions:

250 mL of 0.22 M $K_2Cr_2O_7$
1500 mL of 1.4 M Na_2CO_3
3.15 L of 0.75 M $HClO_4$
857 mL of 0.66 M $CuSO_4 \cdot 5\ H_2O$
50. mL of 0.25 M NaS_2O_3

Do your calculations here.

(5) Using the relationship for dilutions, calculate how much (mL) concentrated reagent is necessary for the following solutions.

$$(\text{volume}_1)(M_1) = (\text{volume}_2)(M_2)$$

15 M NH_4OH to prepare 2.0 L of 3.0 M NH_4OH
14.6 M H_3PO_4 to prepare 300. mL of 0.15 M H_3PO_4
12 M HCl to prepare 6.0 L of 0.10 M HCl
18 M H_2SO_4 to prepare 500 mL of 0.45 M H_2SO_4

Do your calculations here.

11. Water from the Great Salt Lake can have a salt concentration as high 3.42 M (expressed as NaCl). High mountain spring water has a very low salt concentration, 0.00171 M. Potable, or drinkable, water has a maximum recommended level of 0.0171 M or 10 times that of spring water. What is the maximum volume of drinkable water that can be prepared from 500. mL of Great Salt Lake water using spring water for dilution? Assume the salt concentration from the spring water is negligible.

   Do your calculations here.

12. Lead ion can be precipitated out of solution according to the following reaction.

    $$Pb^{2+}(aq) + Na_2CrO_4(aq) \rightarrow PbCrO_4(s) + 2\ Na^+(aq)$$

    What mass of Na_2CrO_4 should be added to 10 L of solution that contains 2.78 g/L of Pb^{2+} ion?

    Do your calculations here.

13. Colligative properties of solutions depend only on the (1) _____ of solute particles present. Therefore, 1 mole of sugar and 1 mole of alcohol, neither of which is ionic, will depress the freezing point or elevate the (2) _____ of a fixed amount of water by an (3) **equal/unequal** amount. These properties are usually expressed on the basis of a fixed amount of solvent, which is (4) _____ , whether water or some other solvent. Other colligative properties include (5) _____ , in addition to freezing point depression and boiling point elevation. A mole of an ionic

substance such as NaCl will lower the freezing point of a solution (6) _____ as much as a nonionic material since (7) _____ ions are produced for each mole of NaCl. A mole of $CaCl_2$ will lower the freezing point of water (8) _____ times as much as a mole of sugar or urea. For un-ionized and nonvolatile substances, molecular masses can be determined from either of two colligative properties, namely, (9) _____ and (10) _____ .

14. (1) How much ethylene glycol, $C_2H_4(OH)_2$, per kilogram of water is needed to lower the freezing from 0°C to −20°C?

   The value for K_f is 1.86°C/mol/kg. The equation is:

   $$\Delta t_f = K_f \times \frac{\text{grams solute}}{\text{molar mass solute}} \times \frac{1}{\text{kg solvent}}$$

   Rearranging the equation to solve for grams of solute gives us

   $$\text{g solute} = \frac{\Delta t_f \times \text{molar mass} \times \text{kg solvent}}{K_f}$$

   First determine the molar mass of ethylene glycol and then make the necessary substitutions in order to solve the equation.

   Do your calculations here.

(2) A sample of an organic compound having a mass of 1.50 g lowered the freezing point of 20.0 g of benzene by 2.75°C. The K_f of benzene is 5.1°C/mol/kg. Calculate the molar mass of the compound. The equation is:

$$\Delta t_f = K_f \times \frac{\text{grams solute}}{\text{molar mass solute}} \times \frac{1}{\text{kg solvent}}$$

Rearranging the equation to solve for molar mass gives us

$$\text{molar mass} = \frac{K_f \times \text{g solute}}{\Delta t_f \times \text{kg solvent}}$$

Make the necessary substitutions and calculate the molar mass of the unknown compound.

Do your calculations here.

15. Calculate the equivalent masses of the following acids and bases using the periodic table in the text.

   (1) $HClO_4$ _____
   (2) KOH _____
   (3) H_2SO_4 _____
   (4) $Ca(OH)_2$ _____
   (5) H_3PO_4 _____
   (6) H_2CO_3 _____
   (7) $Al(OH)_3$ _____
   (8) $Ba(OH)_2$ _____
   (9) HF _____
   (10) $NaOH$ _____
   (11) HCl _____
   (12) NH_4OH _____

16. Determine the Normality, N, of the following solutions whose concentrations are given as Molarity, M.

   (1) 0.15 M H_3PO_4 _____
   (2) 0.1 M HCl _____
   (3) 0.45 M H_2SO_4 _____
   (4) 14.6 M H_3PO_4 _____
   (5) 18 M $HC_2H_3O_2$ _____
   (6) 0.60 M H_2CO_3 _____
   (7) 15 M NH_4OH _____
   (8) 0.70 M $Ca(OH)_2$ _____
   (9) 3.2 M $Al(OH)_3$ _____
   (10) 0.1 M KOH _____
   (11) 2.2 M $NaOH$ _____
   (12) 0.25 M $Ba(OH)_2$ _____

17. What volume of 0.125 N KOH is required to neutralize the following acid solutions?

   (1) 33 mL of 0.25 N H_2SO_4
   (2) 16.5 mL of 0.42 N HCl
   (3) 21.7 mL of 0.080 N H_2SO_4
   (4) 10.55 mL of 0.10 N HNO_3

Challenge Problems

18. Cadmium ion, which is a toxic metallic ion, can be precipitated from solution according to the following reaction.

$$Cd^{2+}(aq) + K_3PO_4(aq) \rightarrow Cd_3(PO_4)_2(s) + K^+(aq)$$

Balance the equation and, assuming no side reactions, determine what volume of stock 5.00 M K_3PO_4 solution should be diluted to 250. mL in order to precipitate 4.10 g Cd^{2+} ion that is contained in 600. mL of a waste solution.

Do your calculations here.

19. One gram (1.00 g) of glucose ($C_6H_{12}O_6$) dissolves in 1.10 mL of H_2O. What is the concentration in terms of mass percent and molality? Human blood normally contains approximately 0.090% glucose (mass percent). What is the molality of blood glucose?

Assume 1000 g of blood is equivalent to 1000 g of solvent.

Do your calculations here.

RECAP SECTION

After completing Chapter 14, you have learned many of the skills that a bench chemist or laboratory technician uses every day. These people often work with solutions; therefore, the knowledge of solution preparation and calculations involving solutions is important to them as well as to other scientists and technicians. The matter of proper preparation of solutions is critical in any work situation that deals with chemicals, such as nursing, agriculture, and food technology. There are many good review exercises at the end of the chapter in the text. Do any problems that are assigned and then try a few more. This is a good opportunity to sharpen up your problem-solving ability.

ANSWERS TO QUESTIONS AND SOLUTIONS TO PROBLEMS

1. (1) bleach, sodium hypochlorite, water
    (2) iodine solution, iodine, carbon tetrachloride
    (3) air, oxygen, nitrogen
    (4) coin alloy, nickel, copper
    (5) lake and river water, oxygen, water

2. (1) s (2) s (3) s (4) i (5) i (6) i (7) s
    (8) i (9) s (10) i (11) s (12) s (13) s (14) i
    (15) s (16) s (17) s (18) i (19) s (20) i

3. (1) unsaturated (2) unsaturated (3) saturated (4) supersaturated
    (5) unsaturated (6) unsaturated (7) unsaturated (8) supersaturated
    (9) unsaturated (10) saturated (11) unsaturated (12) unsaturated

4. (1) polar (2) attracted (3) increase (4) decrease
    (5) increases (6) decreases (7) directly (8) double

5. Four factors are particle size of a solute, stirring, temperature, solution concentration.

6. (1) 30.00 g sucrose, 1470 g H_2O
        A 2.000% sucrose solution means that 2.000% of the total solution is sucrose.

        $$\frac{2.000}{100}(1500 \text{ g}) = 30.00 \text{ g sucrose}$$

        If 30.00 g of the total mass is sucrose, then the mass of water is

        $$1500 \text{ g} - 30.00 \text{ g} = 1470 \text{ g } H_2O$$

    (2) 0.5 g NaCl, 524.5 g H_2O
        A 0.1000% solution means that 0.1000% of 525.0 g is salt.

        $$\left(\frac{0.1000}{100}\right)(525.0 \text{ g}) = 0.5250 \text{ g NaCl}$$
        $$= 0.5 \text{ g}$$

The mass of water is 525 g − 0.5 g = 524.5 g H_2O

7. (1) 30 mL of alcohol
You are asked to calculate a volume contained in a solution of a certain concentration; 12% of the total volume is alcohol.

$$\left(\frac{12}{100}\right)(250 \text{ mL}) = 30. \text{ mL alcohol}$$

As a check:

$$\frac{30.}{250}x = 12\% \text{ alcohol}$$

(2) 1.5×10^2 g of NaOH
Of the 375 g of solution, 40.% is NaOH

$$\left(\frac{40}{100}\right)(375 \text{ g}) = 1.5 \times 10^2 \text{ g of NaOH}$$

8. (1) 14 M H_2SO_4
For any problem involving molarity and quantities of chemicals needed to prepare for various solutions, it is wise to work from the definitions.

From the problem, we have 4.2 moles of H_2SO_4 in 300. mL of solution.

$$M = \frac{4.2 \text{ mol}}{300. \text{ mL}} \times \frac{1000 \text{ mL}}{L} = 14 \text{ M}$$

(2) 0.57 M KCl

$$M = \frac{\text{mol of solute}}{\text{L of solution}} = \frac{1.7 \text{ mol}}{3.0 \text{ L}} = 0.57 \text{ M}$$

9. (1) 3.8 g NaF

molar mass of NaF = 22.99 g + 19.00 g = 41.99 g/mol

We need 600 mL of 0.15 M NaF.

$$M = \frac{g}{\text{molar mass} \times L}$$

Rearranging the equation to solve for g

g = M × molar mass × L

$$\text{g of NaF} = 0.15 \text{ M} \times 41.99 \frac{g}{\text{mol}} \times \frac{600. \text{ mL} \times 1 \text{ L}}{1000 \text{ mL}}$$

We should use the complete units for M so that our answer has the correct units of grams

$$\text{g of NaF} = 0.15 \frac{\text{mol}}{L} \times 41.99 \frac{g}{\text{mol}} \times \frac{600. \text{ mL} \times 1 \text{ L}}{1000 \text{ mL}}$$

$$= 3.8 \text{ g NaF (2 significant figures)}$$

(2) 2.56×10^3 g NH_4NO_3

molar mass of $NH_4NO_3 = 14.01$ g $+ (4 \times 1.008$ g$) + 14.01$ g $+ (3 \times 16.00$ g$)$

$$= 80.05 \, \frac{g}{mol}$$

You need 4.00 L of 8.00 M NH_4NO_3

g = M × molar mass × L

$$= 8.00 \, \frac{\cancel{mol}}{\cancel{L}} \times 80.05 \, \frac{g}{\cancel{mol}} \times 4.00 \, \cancel{L}$$

$$= 2.56 \times 10^3 \text{ g } NH_4NO_3 \text{ (3 significant figures)}$$

(3) 22 g $Ca(OH)_2$

molar mass of $Ca(OH)_2 = 40.08$ g $+ (2 \times 16.00$ g$) + (2 \times 1.008$ g$)$

$$= 74.10 \, \frac{g}{mol}$$

You are asked for 420 mL of 0.70 M $Ca(OH)_2$

g = M × molar mass × L

$$= 0.70 \, \frac{\cancel{mol}}{\cancel{L}} \times 74.10 \, \frac{g}{\cancel{mol}} \times \frac{420 \, \cancel{mL} \times 1 \, \cancel{L}}{1000 \, \cancel{mL}}$$

$$= 22 \text{ g } Ca(OH)_2 \text{ (2 significant figures)}$$

(4) 33 g Na_2SO_4

molar mass of $NaSO_4 = (2 \times 22.99$ g$) + 32.06$ g $+ (4 \times 16.00$ g$)$

$$= 142.0 \, \frac{g}{mol}$$

You are asked for 250 mL of 0.94 M Na_2SO_4

g = M × molar mass × L

$$= 0.94 \, \frac{\cancel{mol}}{\cancel{L}} \times 142.0 \, \frac{g}{\cancel{mol}} \times \frac{250 \, \cancel{mL} \times 1 \, \cancel{L}}{1000 \, \cancel{mL}}$$

$$= 33 \text{ g } Na_2SO_4 \text{ (2 significant figures)}$$

(5) 1.3×10^2 g $NaNO_3$

molar mass of $NaNO_3 = 22.99$ g $+ 14.01$ g $+ (3 \times 16.00$ g$)$

$$= 85.00 \, \frac{g}{mol}$$

You are asked for 0.50 L of 3.00 M $NaNO_3$

g = M × molar mass × L

$$= 3.00 \, \frac{\cancel{mol}}{\cancel{L}} \times 85.00 \, \frac{g}{\cancel{mol}} \times 0.50 \, \cancel{L}$$

$$= 1.3 \times 10^2 \text{ g } NaNO_3 \text{ (2 significant figures)}$$

10. (1) 0.857 L or 857 mL

You are asked to solve the problem for a volume in units of L. With this in mind, write down the complete equation.

$$M = \frac{g}{\text{molar mass} \times L}$$

You now have to rearrange the equation so that "L" is in the numerator and isolated. Multiply both sides of the equation by "L".

$$(L)(M) = \frac{g(L)}{\text{molar mass}(L)}$$

Next, cancel the L's of the right side and divide both sides of the equation by M which gives you:

$$\frac{(L)(M)}{(M)} = \frac{g(L)}{\text{molar mass}(M)(L)}$$

$$(L) = \frac{g}{\text{molar mass}(M)}$$

Remember that M is the same as moles per liter and, when you divide by a fraction, the denominator of the fraction comes up to the numerator.

For example: $\frac{2}{\frac{1}{2}} = \frac{(2)(2)}{1} = 4$

In similar fashion we can write our last equation as

$$(L) = \frac{g}{\text{molar mass}\left(\frac{\text{moles}}{\text{liter}}\right)} = \frac{g(L)}{\text{molar mass (moles)}}$$

Determine the molar mass of $KMnO_4$, substitute the data in the equation, cancel units and solve for L.

molar mass of $KMnO_4 = 39.10\ g + 54.94\ g + (4 \times 16.00\ g) = 158.0\ g/\text{mole}$

Therefore,

$$L = \frac{180.\ g\ (L)}{158.0\ \frac{g}{\text{mol}}(1.33\ \text{mol})} = 0.857\ L\ \text{or}\ 857\ mL$$

(2) 15 L H_2SO_4

$$H_2SO_4 = (2 \times 1.008\ g) + 32.06\ g + (4 \times 16.00\ g) = 98.08\ \frac{g}{\text{mol}}$$

$$L = \frac{g}{\text{molar mass} \times M}$$

This equation is solved just like the one above in 10. (1)

$$= \frac{147\ g \times 1\ L}{98.08\ \frac{g}{\text{mol}} \times 0.10\ \text{mol}} = 15\ L$$

(3) 1.3 moles of HCl

For this type of problem, go back to the simple first relationship since you are not asked a question involving grams or molar mass. The defining equation for molarity is

$$M = \frac{mol}{L}$$

Therefore,

$$moles = L \times M$$

This is what you have been asked to find in the above problem.

$$\text{moles of HCl} = 4.0 \, L \times 0.333 \, \frac{mol}{L} = 1.3 \, mol$$

(4) $K_2Cr_2O_7 = 0.055$ mole, $Na_2CO_3 = 2.1$ moles.
$HClO_4 = 2.4$ moles; $CuSO_4 \cdot 5\,H_2O = 0.57$ mole; $Na_2S_2O_3 = 0.013$ mole

$$moles = L \times M$$

$K_2Cr_2O_7$ $\qquad$ $moles = \frac{250 \, mL \times 1\,L}{1000 \, mL} \times 0.22 \, \frac{mol}{L} = 0.055 \, mol$

Na_2CO_3 $\qquad$ $moles = \frac{1500 \, mL \times 1\,L}{1000 \, mL} \times 1.4 \, \frac{mol}{L} = 2.1 \, mol$

$HClO_4$ $\qquad$ $moles = 3.15 \, L \times 0.75 \, \frac{mol}{L} = 2.4 \, mol$

$CuSO_4 \cdot 5\,H_2O$ $\qquad$ $moles = \frac{857 \, mL \times 1\,L}{1000 \, mL} \times 0.66 \, \frac{mol}{L} = 0.57 \, mol$

$Na_2S_2O_3$ $\qquad$ $moles = \frac{50. \, mL \times 1\,L}{1000 \, mL} \times 0.25 \, \frac{mol}{L} = 0.013 \, mol$

(5) $NH_4OH = 400$ mL, $H_3PO_4 = 3.0$ mL, $HCl = 50$ mL, $H_2SO_4 = 13$ mL

In each case, we must solve the relationship for the initial volume (V_1), which is the quantity of the concentrate reagent. Even though we dilute the reagent with water, the number of moles of chemical remains the same; the equation is a simple equality.

For the first problem, we dilute 15 M NH_4OH (ammonium hydroxide) to 2.0 L of 3.0 M NH_4OH.

Our equation is

$$(V_1)(M_1) = (V_2)(M_2)$$
$$V_1 = \frac{(V_2)(M_2)}{M_1}$$

Substituting in the equation

$$V_1 = \frac{(2.0 \text{ L})(3.0 \text{ M})}{(15 \text{ M})}$$

$V_1 = 0.40$ L or 4.0×10^2 mL of 15 M NH_4OH diluted to 2.0 L with water

H_3PO_4 You have to change 300. mL into L before substituting in the equation.

$$V_1 = \frac{(0.300 \text{ L})(0.15 \text{ M})}{(14.6 \text{ M})}$$

$V_1 = 0.0031$ L = 3.1 mL

HCl

$$V_1 = \frac{(6.0 \text{ L})(0.10 \text{ M})}{(12 \text{ M})}$$

$V_1 = 0.050$ L = 50. mL

H_2SO_4 You have to change 500. mL into L first.

$$V_1 = \frac{(0.500 \text{ L})(0.45 \text{ M})}{(18 \text{ M})}$$

$V_1 = 0.013$ L = 13 mL

11. 100. L

This is a dilution problem and we can use the equation

$$V_1 M_1 = V_2 M_2$$

where $V_1 = 500.$ mL $= 0.500$ L
$M_1 = 3.42$ M $= 3.42$ mol/L
$M_2 = 0.0171$ M $= 0.0171$ mol/L
$V_2 = $ Unknown final volume

Substituting into the equation and solving for V_2

$$V_2 = \frac{(0.500 \text{ L})(3.42 \text{ mol/L})}{0.0171 \text{ mol/L}}$$

$= 100.$ L

12. 21.7 g Na_2CrO_4

The reaction equation is balanced as written and says that 1 mole of Pb^{2+} ion reacts with 1 mole of CrO_4^{2-} ion. So, we need to calculate how many moles of Pb^{2+} ion we have and from this calculate the mass of Na_2CrO_4 needed.

moles of Pb^{2+} = (mass Pb^{2+}/molar mass Pb^{2+}) × 10 L

$$= \frac{2.78 \text{ g/L}}{207.2 \text{ g/mol}} \times 10.0 \text{ L} = 0.134 \text{ mol}$$

In order to precipitate 0.134 moles of Pb^{2+} ion, we must add an equal number of moles of Na_2CrO_4 which we can determine as follows:

molar mass of Na_2CrO_4 = 45.98 g + 52.00 g + 64.00 g = 162 g/mol
mass Na_2CrO_4 = (0.134 mol)(162 g/mol) = 21.7 g

13. (1) number
    (2) boiling point
    (3) equal
    (4) 1 kg
    (5) vapor pressure lowering
    (6) about twice
    (7) 2 moles
    (8) about three
    (9) freezing point depression
    (10) boiling point elevation

14. (1) 6.67×10^2 g

    The problem states that the temperature differential is from 0.0°C to −20.0°C or a Δt of 20°. In addition, we have the value for K_f for water and the fact that we are dealing with 1 kilogram of solvent. We need to determine the molar mass of ethylene glycol and then use the equation given.

    molar mass of $C_2H_4(OH)_2$ = 2 × 12.01 g = 24.02 g
    4 × 1.008 g = 4.032 g
    2 × 16.00 g = 32.00 g
    2 × 1.008 g = 2.016 g
    62.07 g/mol

    $$\text{g solute} = \frac{\Delta t_f \times \text{molar mass} \times \text{kg solvent}}{K_f}$$

    Making the appropriate substitutions:

    $$= \frac{20.0°C \times 62.07 \text{ g/mol} \times 1 \text{ kg solvent}}{1.86 °C/\text{mol}/\text{kg}}$$

    $= 6.67 \times 10^2$ g of ethylene glycol

    (2) 140 g/mol

    We are asked in this problem to determine the molar mass of a compound from freezing point depression data. In order to solve the equation, we need to express the amount of solvent in units of kilograms and then make the necessary substitutions of the other values.

    20.0 g of benzene = 0.0200 kg

    K_f = 5.1°C/mol/kg

    Δt_f = 2.75°C

    grams of solute = 1.50 g

    $$\text{molar mass} = \frac{K_f \times \text{g solute}}{\Delta t_f \times \text{kg slovent}}$$

    $$= \frac{5.1°C \times 1.50 \text{ g}}{\frac{\text{mol}}{\text{kg}} \times 2.75°C \times 0.0200 \text{ kg}}$$

    = 140 g/mol

15. (1) $HClO_4$ 100.5 g/equiv   (7) $Al(OH)_3$ 26.00 g/equiv
    (2) KOH   56.11 g/equiv   (8) $Ba(OH)_2$ 85.66 g/equiv
    (3) H_2SO_4 49.04 g/equiv   (9) HF   20.01 g/equiv
    (4) $Ca(OH)_2$ 37.05 g/equiv   (10) NaOH   40.00 g/equiv
    (5) H_3PO_4 32.66 g/equiv   (11) HCl   36.46 g/equiv
    (6) H_2CO_3 31.01 g/equiv   (12) NH_4OH 35.05 g/equiv

16. (1) 0.45 N H_3PO_4 (7) 15 N NH_4OH
    (2) 0.1 N HCl   (8) 1.4 N $Ca(OH)_2$
    (3) 0.90 N H_2SO_4 (9) 9.6 N $Al(OH)_3$
    (4) 43.8 N H_3PO_4 (10) 0.1 N KOH
    (5) 18 N $HC_2H_3O_2$ (11) 2.2 N NaOH
    (6) 1.2 N H_2CO_3 (12) 0.50 N $Ba(OH)_2$

17. (1) 66 mL of 1.25 N KOH
    (2) 55 mL
    (3) 14 mL
    (4) 8.4 mL

18. Balanced equation is $3\ Cd^{2+}(aq) + 2\ K_3PO_4(aq) \rightarrow Cd_3(PO_4)_2(s) + 6\ K^+(aq)$. Volume of 5.00 M of K_3PO_4 required is 4.86 mL. The equation states that 3 moles of Cd^{2+} require 2 moles of K_3PO_4 to precipitate. The number of moles of Cd^{2+} ion present is equal to

$$\frac{4.10\ g\ Cd^{2+}}{112.4\ g/mol} = 0.0365\ mol\ Cd^{2+}$$

We will need the following number of moles of K_3PO_4.

$$\text{Moles } K_3PO_4 = (0.0365\ mol\ Cd^{2+}) \frac{2\ mol\ K_3PO_4}{3\ mol\ Cd^{2+}} = 0.0243\ mol$$

We next use dilution equation $V_1M_1 = V_2M_2$ where V_1 = unknown volume of stock 5.00 M solution and M_1 = 5.00 M.

Since 250. mL of the diluted K_3PO_4 is added to 600. mL of the Cd^{2+} solution, the final volume (V_2) will equal 850. mL (0.850 L).

$$V_2 = 250.\ mL + 600.\ mL = 0.850\ L$$

$$M_2 = \frac{0.0243\ mol}{0.850\ L} = 0.0286\ M$$

$$V_1 = \frac{(0.850\ L)(0.0286\ M)}{5.00\ M}$$

$$= 0.00486\ L = 4.86\ mL$$

19. Mass percent of glucose solution is 47.6%; molality is 5.05 m. Blood glucose molality is 0.50 m. The molar mass of glucose is 180.2 g/mol. A solution of 1.00 g glucose is 1.10 mL of H_2O is equal to a mass percent concentration of

$$\frac{1.00 \text{ g glucose}}{(1.10 \text{ g}) + (1.00 \text{ g})} \times 100 = 47.6\%$$

We can dissolve $\frac{1000.}{1.10}$ g of glucose in 1 kg of water.

This amount is equal to 909 g of glucose or $\frac{909 \text{ g}}{180.2 \text{ g/mol}} = 5.04$ mol

The molality is therefore 5.04 m.

Human blood has a concentration of 0.090% or 0.90 g per 1000 g of blood. Using the assumption that 1000 g of blood is equivalent to 1000 g of solvent, the molality can be calculated as follows:

$$\frac{0.90 \text{ g}}{180.2 \text{ g/mol}} = 0.0050 \text{ mol}$$

$$\text{molality} = \frac{0.0050 \text{ mol}}{1 \text{ kg solvent}} = 0.0050 \ m$$

CROSSWORD PUZZLE 1

Across

2. A mixture without a uniform composition.
6. Yellow solid nonmetallic element (atomic number 16).
7. An element that is ductile and malleable is a _____.
10. One shouldn't do crossword puzzles with a _____.
11. Binary compounds with a metallic element have names ending in _____.
12. Number of protons in hydrogen nucleus.
13. Chemical symbol for tantalum.
15. Chemical symbol for cobalt.
16. Elements whose d and f orbitals are being filled.
17. Formula for nitrogen oxide.
18. Light energy is sometimes considered to exist as a _____.
20. Chemical symbol for element number 95.
22. A pair of electrons forms a chemical _____ between two atoms.
24. Chemical symbol for osmium.
26. A positively charged ion.
27. Chemical symbol for element number 10.

Down

1. The answer to a mathematical problem is often called the _____.
3. Oxygen is an example of an _____, the basic building blocks of matter.
4. Nitrogen exists in what physical state as the uncombined element.
5. Chemical symbol for first element in period number 3.
8. Heat and light are forms of _____.
9. Element number 54.
10. Chemical symbol for chemical number 84.
13. The element whose chemical symbol is Sn.
14. The smallest indivisible unit of matter.
15. When elements combine chemically, they form a new substance called a _____.
19. A negatively charged ion is called an _____.
21. One Avogadro's Number of a substance dissolved in enough water to give 1 liter of solution is a 1 _____ solution.
23. Chemical symbol for element number 28.
25. Chemical symbol for element number 34.

Answers are found at the end of Chapter 20.

CHAPTER FIFTEEN

Acids, Bases, and Salts

SELF-EVALUATION SECTION

1. There are several different definitions of acids and bases according to various theories. Match the chemist's name with the appropriate phrase:

   (1) Arrhenius   a. Acid is proton donor

   (2) Brønsted-Lowry   b. Acid is electron pair acceptor

   (3) Lewis   c. Acids are hydrogen-containing substances that dissociate to produce H^+

2. Identify the conjugate acid-base pairs in the following equations.

   (1) $H_2BO_3^- + H_3O^+ \rightarrow H_3BO_3 + H_2O$

   (2) $NH_3 + HBr \rightarrow NH_4^+ + Br^-$

   (3) $HSO_4^- + H_3O^+ \rightarrow H_2SO_4 + H_2O$

   (4) $HC_2H_3O_2 + H_2O \rightarrow H_3O^+ + C_2H_3O_2^-$

   (5) $NH_4^+ + H_2O \rightarrow NH_3 + H_3O^+$

3. Complete and balance the following reactions between HCl and the chemicals given:

   (1) HCl and NH_4OH

   (2) HCl and Zn metal

   (3) HCl and CaO

   (4) HCl and Na_2CO_3

4. Complete the equation for the reaction of HCl and $Al(OH)_3$, an amphoteric hydroxide.

   $HCl + Al(OH)_3 \rightarrow$

5. Neutralization reactions. Identify the reactants as either *acids* or *bases*.

   (1) $NH_4OH + HClO_3 \rightarrow NH_4ClO_3 + H_2O$

   _____ _____

(2) $Al(OH)_3 + H_3PO_4 \rightarrow AlPO_4 + 3\ H_2O$

_____ _____

(3) $H_2SO_4 + Ca(OH)_2 \rightarrow CaSO_4 + 2\ H_2O$

_____ _____

Dissociation reactions. Identify the acids and bases in each reaction according to the Brønsted-Lowry theory. Examine both the reactants and the products.

(4) $NH_4^+ + H_2O \rightarrow NH_3 + H_3O^+$

_____ _____ _____ _____

(5) $C_2H_3O_2^- + H_2O \rightarrow HC_2H_3O_2 + OH^-$

_____ _____ _____ _____

(6) $Al^{3+} + H_2O \rightarrow Al(OH)_3^{2+} + H^+$

_____ _____ _____ _____

6. In which of the following situations does the solution contain an elctrolyte (E) or nonelectrolyte (NE)?

   (1) The solution conducts an electric current.   _____

   (2) The solution contains only electrically neutral molecules.   _____

   (3) The solution contains electrically charged ions.   _____

   (4) The solution contains an acid, such as H_2SO_4.   _____

   (5) The solution contains gaseous oxygen.   _____

   (6) The solution is a nonconductor of electric current.   _____

   (7) The solution contains molecules, such as acetic acid, which have reacted with water to produce ions.   _____

7. Identify the following reactions as either dissociation (D) or ionization (I).

   (1) $NaCl \xrightarrow{H_2O} Na^+ + Cl^-$ _____

   (2) $HC_2H_3O_2 + H_2O \rightarrow H_3O^+ + C_2H_3O_2^-$ _____

   (3) $NH_3 + H_2O \rightarrow NH_4^+ + OH^-$ _____

   (4) $Ca(OH)_2 \xrightarrow{H_2O} Ca^{2+} + 2\ OH^-$ _____

   (5) $HCl + H_2O \rightarrow H_3O^+ + Cl^-$ _____

8. Calculate the molarity of the ions in each of the salt solutions listed below. Consider each salt to be 100% dissociated.

   (1) 0.15 M $CaCl_2$ (4) 1.4 M LiI

   (2) 0.75 M $(NH_4)_2SO_4$ (5) 0.022 M $Al_2(SO_4)_3$

   (3) 3.3 M Na_3PO_4

9. Identify the following compounds as strong electrolytes (s) or weak electrolytes (w).

(1) HF
(2) $MgNO_3$
(3) NaOH
(4) $HC_2H_3O_2$
(5) $CaCl_2$
(6) HCl
(7) HClO
(8) H_3BO_3
(9) $HClO_4$
(10) NH_4OH

10. Rank these common solutions from strong acid to strong base. Indicate which solutions are acid, close to neutral, and basic using Figure 15.3.

	Solution	pH	Rank in order from acid to base
(1)	0.1 M HCl	1.0	
(2)	Blood	7.4	
(3)	Lemon juice	2.3	
(4)	Drain cleaner	12.1	
(5)	Vinegar	2.8	
(6)	0.1 M $HC_2H_3O_2$	2.9	
(7)	Milk	6.6	
(8)	Ammonium cleaner	8.2	
(9)	Carbonated water	3.0	
(10)	Tomato juice	4.1	

11. Let's do a little bit more with pH. The mathematical expression for pH indicates that we use the hydrogen ion concentration. For example, when $[H^+] = 1 \times 10^{-3}$ mol/L, the pH is 3. When the $[H^+] = 1 \times 10^{-9}$ mol/L, the pH is 9.

As you can see, we derive the pH value from the exponent of 10 when the number in front is 1. When the number in front of the exponent is between 1 and 10, the pH is between the power of 10 given and the next lower power.

As an example

$[H^+] = 6 \times 10^{-7}$ mol/L   pH is between 7 and 6
$[H^+] = 5.4 \times 10^{-2}$ mol/L   pH is between 2 and 1

(1) What is the pH of a solution whose $[H^+] = 1 \times 10^{-10}$ mol/L?

pH = _____

(2) What is the pH of a solution whose $[H^+] = 1 \times 10^{-7}$ mol/L?

pH = _____

(3) What is the pH of a solution whose $[H^+] = 4 \times 10^{-8}$ mol/L?

pH = _____

(4) What is the pH of a solution whose $[H^+] = 3.3 \times 10^{-5}$ mol/L?

pH = _____

12. Indicate whether the following equations are *un-ionized* equations (U), *total ionic* equations (T) or *net ionic* equations (N).

   (1) $2\ OH^-(aq) + Mn^{2+}(aq) \rightarrow Mn(OH)_2(s)$ _____

   (2) $2\ HCl(aq) + Ca(s) \rightarrow H_2(g) + CaCl_2(aq)$ _____

   (3) $3\ NH_4OH(aq) + FeCl_3(aq) \rightarrow Fe(OH)_3(s) + 3\ NH_4Cl(aq)$ _____

   (4) $2\ H^+(aq) + SO_4^{2-}(aq) + Mg(s) \rightarrow H_2(g) + Mg^{2+}(aq) + SO_4^{2-}(aq)$ _____

   (5) $2\ H^+(aq) + 2\ Cl^-(aq) + Na_2O(s) \rightarrow 2\ Na^+(aq) + 2\ Cl^-(aq) + H_2O(l)$ _____

   (6) $2\ OH^-(aq) + Zn(s) \rightarrow ZnO_2^{2-}(aq) + H_2(g)$ _____

   (7) $H^+ + OH^- \rightarrow H_2O$ _____

   (8) $Zn(OH)_2(s) + 2\ H^+(aq) + 2\ Cl^-(aq) \rightarrow$
   $\qquad Zn^{2+}(aq) + 2\ Cl^-(aq) + 2\ H_2O(l)$ _____

   (9) $2\ H^+(aq) + CO_3^{2-}(aq) \rightarrow H_2O(l) + CO_2(g)$ _____

   (10) $H_2SO_4(aq) + MgCO_3(aq) \rightarrow MgSO_4(aq) + H_2O(l) + CO_2(g)$ _____

13. Write balanced *un-ionized, total ionic,* and *net ionic* equations for the following two equations written in words.

   (1) Barium chloride solution plus silver nitrate solution react to form insoluble silver chloride plus barium nitrate solution.

   Un-ionized

   Total ionic

   Net ionic

   (2) Solid iron plus copper(II) sulfate solution react to form solid copper plus iron(II) sulfate solution.

   Un-ionized

   Total ionic

— 148 —

Net ionic

14. 35 mL of 0.20 M HCl is required to titrate 50. mL of Ca(OH)$_2$ solution. What is the molarity of the base?

   Do your calculations here.

15. A 25.00 mL sample of Ca(OH)$_2$ solution required 21.35 mL of 0.3675 N H$_2$SO$_4$ for complete neutralization. What is the normality and the molarity of the calcium hydroxide? What is the molarity of the sulfuric acid?

   Do your calculations here.

Challenge Problems

16. Identify the conjugate acid-base pairs in the three step ionization of phosphoric acid, H_3PO_4. Write each equation and indicate the acid-base pairs.

    Do your calculations here.

17. How many grams, theoretically, of $BaSO_4$ will be precipitated when 35 mL of 0.15 M H_2SO_4 is added to 18 mL of 0.43 M $Ba(OH)_2$? What is the limiting reactant? If the actual yield is 1.0 g $BaSO_4$, what is the percent yield?

    Do your calculations here.

18. You are an analyst working on product quality for a vinegar company. Your work this morning involves 10.00 mL samples of various batches of wine vinegar. The data table for the four titrations looks like this:

Sample volume:	10.00 mL			
	Batch A	B	C	D
Volume of 0.2500 N NaOH	33.05 mL	34.75 mL	32.80 mL	34.35 mL

What is the average normality of acid in the vinegar? If the acidity is due to acetic acid ($HC_2H_3O_2$) what is the average molarity of the vinegar? Given that the density of the vinegar is 1.005 g/mL, what is the average mass percent concentration?

RECAP SECTION

Chapter 15 is an excellent chapter to be related to your laboratory experiments. You will be working with acids and bases, doing titrations, using the pH system, and examining the properties of electrolytes, nonelectrolytes, and colloids. Each of these topics is an important part of Chapter 15, and laboratory experiments will help clarify what was covered in lectures and homework problems. The last section on equation writing will be carried over to Chapter 17 when you examine oxidation-reduction reactions and try your hand at balancing all types of reaction equations.

ANSWERS TO QUESTIONS AND SOLUTIONS TO PROBLEMS

1. (1) c (2) a (3) b
2. (1) $H_2BO_3^-$ base, H_3BO_3 acid; H_3O^+ acid, H_2O base
   (2) NH_3 base, NH_4^+ acid; HBr acid, Br^- base
   (3) HSO_4^- base, H_2SO_4 acid; H_3O^+ acid, H_2O base
   (4) $HC_2H_3O_2$ acid, $C_2H_3O_2^-$ base; H_2O base, H_3O^+ acid
   (5) NH_4^+ acid, NH_3 base; H_2O base, H_3O^+ acid
3. (1) $HCl(aq) + NH_4OH(aq) \rightarrow NH_4Cl(aq) + H_2O(l)$
   (2) $2\ HCl(aq) + Zn(s) \rightarrow ZnCl_2(aq) + H_2(g)$
   (3) $2\ HCl(aq) + CaO(s) \rightarrow CaCl_2(aq) + H_2O(l)$
   (4) $2\ HCl(aq) + Na_2CO_3(aq) \rightarrow 2\ NaCl(aq) + H_2O + CO_2(g)$
4. $3\ HCl(aq) + Al(OH)_3(s) \rightarrow AlCl_3(aq) + 3\ H_2O(l)$
5. (1) NH_4OH = base $HClO_3$ = acid
   (2) $Al(OH)_3$ = base H_3PO_4 = acid
   (3) H_2SO_4 = acid $Ca(OH)_2$ = base
   (4) NH_4^+ = acid; H_2O = base; NH_3 = base; H_3O^+ = acid
   (5) $C_2H_3O_2^-$ = base H_2O = acid; $HC_2H_3O_2$ = acid; OH^- = base
   (6) Al^{3+} = acid; H_2O = base; $Al(OH)^{2+}$ = base; H^+ = acid
6. (1) E (2) NE (3) E (4) E (5) NE (6) NE (7) E
7. (1) Dissociation (D). Ions are present in the crystalline salt, and the water molecules act only as the separation agent. Molten salt as well as a solution of salt conducts electricity.

   (2) Ionization (I). Water molecules are necessary to form ions in reaction with $HC_2H_3O_2$. Pure $HC_2H_3O_2$ is a very weak electrolyte.

(3) Ionization (I). When ammonia gas dissolves in water, it undergoes reaction with water molecules to form a solution we call ammonium hydroxide (NH_4OH).

(4) Dissociation (D).

(5) Ionization (I). The bond in $HCl(g)$ is predominately covalent.

8. (1) 0.15 M Ca^{2+}, 0.30 Cl^-
   (2) 1.5 M NH_4^+, 0.75 M SO_4^{2-}
   (3) 9.9 M Na^+, 3.3 M PO_4^{3-}
   (4) 1.4 M Li^+, 1.4 M I^-
   (5) 0.044 M Al^{3+}, 0.066 M SO_4^{2-}

9. (1) w   (2) s   (3) s   (4) w   (5) s
   (6) s   (7) w   (8) w   (9) s   (10) w

10. (1)   (3)   (5)   (6)   (9)   (10)   (7)   (2)   (8)   (4)
    The first six (in order above) are acids, next two close to neutral, and the last two are basic.

11. (1) 10   (2) 7   (3) Between 8 and 7   (4) Between 5 and 4

12. (1) N   (2) U   (3) U   (4) T   (5) T
    (6) N   (7) U   (8) T   (9) N   (10) U

13. (1) Un-ionized (U) $BaCl(aq) + 2\ AgNO_3(aq) \rightarrow 2\ AgCl(s) + Ba(NO_3)_2(aq)$
    Total ionic (T) $Ba^{2+}(aq) + 2\ Cl^-(aq) + 2Ag^+(aq) + 2\ NO_3^-(aq) \rightarrow$
    $\qquad 2\ AgCl(s) + Ba^{2+}(aq) + 2NO_3^-(aq)$
    Net ionic (N) $Ag^+(aq) + Cl^-(aq) \rightarrow AgCl(s)$

    (2) Un-ionized $Fe(s) + CuSO_4(aq) \rightarrow Cu(s) + FeSO_4(aq)$
    Total ionic $Fe(s) + Cu^{2+}(aq) \rightarrow Cu(s) + Fe^{2+}(aq) + SO_4^{2-}(aq)$
    Net ionic $Fe(s) + Cu^{2+}(aq) \rightarrow Cu(s) + Fe^{2+}(aq)$

14. 0.070 M. Equation is $2\ HCl(aq) + Ca(OH)_2(aq) \rightarrow CaCl_2(aq) + 2\ H_2O(l)$. The number of moles of acid used is twice the number of moles of base. Therefore, if we calculate the number of moles of acid used we can use the mole-ratio method to determine the number of moles of $Ca(OH)_2$ in the 50. mL volume, and thus the concentration or molarity.

$$\text{number of moles of HCl} = 0.035\ \cancel{L} \times \frac{0.20\ \text{mol HCl}}{\cancel{L}}$$
$$= 0.0070\ \text{mol}$$

Using the mole ratio to determine the number of moles of $Ca(OH)_2$

$$0.0070\ \cancel{\text{mol HCl}} \times \frac{1\ \text{mol Ca(OH)}_2}{2\ \cancel{\text{mol HCl}}} = 0.0035\ \text{mol Ca(OH)}_2$$

Therefore, 0.0035 mol of $Ca(OH)_2$ was present in 50 mL of $Ca(OH)_2$ solution.

Molarity of $Ca(OH)_2$

$$M = \frac{\text{mol}}{L} = \frac{0.0035\ \text{mol Ca(OH)}_2}{0.050\ L} = 0.070\ M\ Ca(OH)_2$$

15. The normality of the $Ca(OH)_2$ can be calculated from

$$V_A N_A = V_B N_B$$

substituting the data and solving for N_B

$$N_B = \frac{21.35 \text{ mL} \times 0.3675 \text{ N}}{25.00 \text{ mL}} = 0.3138 \text{ N } Ca(OH)_2$$

Since $Ca(OH)_2$ supplies 2 equivalents of OH^- per mol the conversion to molarity is

$$\frac{0.3138 \text{ equiv } Ca(OH)_2}{L} \times \frac{1 \text{ mol } H_2SO_4}{2 \text{ equiv } Ca(OH)_2} = 0.1569 \text{ mol/L}$$

The $Ca(OH)_2$ solution is 0.1569 M.

The molarity of the sulfuric acid can be found in a similar manner,

$$\frac{0.3675 \text{ equiv } H_2SO_4}{L} \times \frac{1 \text{ mol } H_2SO_4}{2 \text{ equiv } H_2SO_4} = 0.1838 \text{ mol/L}$$

The H_2SO_4 solution is 0.1838 M.

16. (1) $H_3PO_4 + H_2O \rightarrow H_3O^+ + H_2PO_4^-$
    (2) $H_2PO_4^- + H_2O \rightarrow H_3O^+ + HPO_4^{2-}$
    (3) $HPO_4^{2-} + H_2O \rightarrow H_3O^+ + PO_4^{3-}$

In each case H_3O^+ and H_2O are a conjugate acid-base pair. For reaction (1) H_3PO_4 is the acid and $H_2PO_4^-$ is the base. For reaction (2) $H_2PO_4^-$ is the acid and HPO_4^{2-} is the base while in (3) HPO_4^{2-} is the acid and PO_4^{3-} is the base.

17. H_2SO_4 is limiting, theoretical yield is 1.2 g $BaSO_4$, percent yield is 83%.

Net ionic equation is $Ba^{2+} + SO_4^{2-} \rightarrow BaSO_4(s)$

Number of moles $Ba^{2+} = \left(0.43 \frac{mol}{L}\right)(0.018 \text{ L}) = 0.0077$ mol

Number of moles $SO_4^{2-} = \left(0.15 \frac{mol}{L}\right)(0.035 \text{ L}) = 0.0053$ mol

Molar mass $BaSO_4 = 233 \frac{g}{mol}$

Amount $BaSO_4 = \left(233 \frac{g}{mol}\right)(0.0053 \text{ mol}) = 1.2$ g

Percent yield $= \frac{\text{Actual yield}}{\text{Theoretical yield}} \times 100 = \frac{1.0 \text{ g}}{1.2} \times = 83\%$

18. The normality of each batch can be calculated from the titration data and the equation

$$N_A V_A = N_B V_B$$

$$N_A = \frac{N_B V_B}{V_A}$$

Substituting data for each batch into the equation we find that

	Batch A	B	C	D
Normality (N_A)	0.8263 N	0.8688 N	0.8200 N	0.8588 N

The average normality of the acid is: 0.8435 N

Since acetic acid ($HC_2H_3O_2$) has one equivalent of H^+ ion per mole, the average molarity will be the same value as the normality.

$$\text{Molarity of } HC_2H_3O_2 = 0.8435 \text{ M}$$

The number of grams of acetic acid can be calculated from the molarity relationship.

$$0.8435 \, \frac{\text{mol}}{\text{L}} \times 60.06 \, \frac{\text{g}}{\text{mol}} = 50.66 \text{ g/L}$$

If the density of the vinegar is 1.005 g/mL then 1000. mL of vinegar will have a mass of:

$$1.005 \text{ g/mL} \times 1000 \text{ mL} = 1005 \text{ g}$$

The mass percent calculation is

$$\frac{50.66 \text{ g}}{1005 \text{ g}} \times 100 = 5.041\% \, HC_2H_3O_2$$

CHAPTER SIXTEEN

Chemical Equilibrium

SELF-EVALUATION SECTION

Most of the examples and problems that we examine for Chapter 16 will have close counterparts in the text material. You should refer to the examples in the text as needed.

1. Write the following word equations for reversible systems in chemical equation form.

   (1) Two moles of sulfur dioxide gas plus one mole of oxygen gas react to produce two moles of sulfur trioxide gas.

   _____

   (2) Four moles of ammonia gas plus five moles of oxygen gas react to produce four moles of nitrogen oxide gas plus six moles of water vapor.

   _____

   (3) One mole of nitrogen gas plus three moles of hydrogen gas react to produce two moles of ammonia gas plus heat.

   _____

   (4) Three moles of oxygen gas plus heat react to produce two moles of ozone gas. One molecule of ozone consists of three atoms of oxygen.

   _____

   (5) One mole of carbon in the form of coke plus one mole of carbon dioxide gas plus heat react to produce two moles of carbon monoxide gas.

   _____

2. Using the Principle of Le Chatelier, predict the effect of changing various conditions on the following equilibrium systems.

   (1) $2\,SO_2(g) + O_2(g) \rightleftarrows 2\,SO_3(g)$

   Which direction, right or left, will the equilibrium be shifted if:

   a. the amount of SO_2 is decreased? _____

   b. the pressure is increased? _____

   c. a catalyst is added? _____

— 157 —

(2) $3\,O_2(g) + \text{heat} \rightleftharpoons 2\,O_3(g)$

Which direction will the equilibrium be shifted if:

a. the amount of ozone is decreased? _____

b. the amount of oxygen is increased? _____

c. the reaction is cooled? _____

d. the pressure is decreased? _____

(3) $C(s) + CO_2(g) \rightleftharpoons 2\,CO(g)$

Which direction will the equilibrium be shifted if:

a. the amount of carbon is increased? _____

b. the reaction mixture is heated? _____

c. the amount of CO is increased? _____

d. the amount of CO_2 is increased? _____

e. the pressure is increased? _____

3. Write the equilibrium constant expression for the following reactions.

(1) $2\,SO_2(g) + O_2(g) \rightleftharpoons 2\,SO_3(g)$

(2) $3\,O_2(g) \overset{\Delta}{\rightleftharpoons} 2\,O_3(g)$

(3) $C(s) + CO_2(g) \overset{\Delta}{\rightleftharpoons} 2\,CO(g)$

(4) $HNO_2 \rightleftharpoons H^+ + NO_2^-$

(5) $Ag^+ + 2\,NH_3 \rightleftharpoons Ag(NH_3)_2^+$

(6) $Al^{3+} + OH^- \rightleftharpoons Al(OH)^{2+}$

(7) $PbSO_4(s) \rightleftharpoons Pb^{2+}(aq) + SO_4^{2-}(aq)$

(8) $CH_4(g) + 2\,O_2(g) \rightleftharpoons CO_2(g) + 2\,H_2O(g)$

(9) $4\,HCl(g) + O_2(g) \rightleftharpoons 2\,H_2O(g) + 2\,Cl_2(g)$

(10) $Cl_2(g) + H_2O(l) \rightleftharpoons HClO(aq) + HCl(aq)$

Write your answers here.

4. What is the value of K_{eq} for the reaction shown if a 1.0 L flask originally contains 0.5 moles of $SO_3(g)$ at 832°C and, at equilibrium, 50 percent of the SO_3 gas has decomposed?

$$2\ SO_3(g) \rightleftarrows 2\ SO_2(g) + O_2(g)$$

5. (1) What is the concentration of hydrogen ions, H^+, in a 0.25 M solution of nitrous acid, HNO_2? $K_a = 4.5 \times 10^{-4}$

Do your calculations here.

(2) What is the concentration of hydrogen ions, H^+, in a 0.15 M solution of butyric acid, $CH_3(CH_2)_2COOH$? $K_a = 1.52 \times 10^{-5}$

Do your calculations here.

6. (1) What is the H^+ ion concentration and the OH^- ion concentration in a 0.1 M HBr solution? Assume the HBr is 100% ionized. $K_w = 1 \times 10^{-14}$

Do your calculations here.

(2) What is the H^+ ion concentration and the OH^- ion concentration in a solution of pH 5?

Do your calculations here.

(3) What is the pH of a 1×10^{-4} M KOH solution? Assume that the KOH is 100% ionized.

Do your calculations here.

(4) Let's review in a short drill the relationships between pH, pOH, H^+, and OH^- concentrations.

    a.   The pH of a solution is 4.

            What is the pOH? _____

            What is the H^+ concentration? _____

            What is the OH^- concentration? _____

b. The pOH of a solution is 8.

What is the OH^- concentration? _____

What is the pH? _____

What is the H^+ concentration? _____

c. The H^+ concentration is 1×10^{-11} mol/L.

What is the pH? _____

What is the pOH? _____

What is the OH^- concentration? _____

(5) Fill in the blank space with a correct word or term.

A solution that consists of a weak acid or base and the corresponding salt, and which resists pH changes when diluted or when small amounts of strong acid or base are added is called a (1) _____ solution. In the acetic acid-sodium acetate buffer system, added H^+ (from HCl) reacts with the (2) _____ ion to form un-ionized (3) _____ . If hydroxide ion (OH^-) from NaOH were added, it would react with the (4) _____ molecule to form two products (5) _____ .

(6) Look at the compounds listed below. Match up the pairs that would be used for various buffer solutions.

a. $HC_2H_3O_2$ d. $H_2PO_4^-$

b. Na_2HPO_4 e. $NaC_2H_3O_2$

c. H_2CO_3 f. $NaHCO_3$

7. The phosphate buffer system, which is important to the acid-base equilibrium of red blood cells, can be described by the following equations.

$$H_2PO_4^- \rightleftharpoons H^+ + HPO_4^{2-}$$
$$Na_2HPO_4 \rightleftharpoons 2\,Na^+ + HPO_4^{2-}$$

Using chemical equations, describe what happens to the phosphate buffer system when (a) acid is added or (b) base is added.

Write your answers here.

8. (1) Lead carbonate, $PbCO_3$, is a slightly soluble salt that dissociates according to the following reaction:

$$PbCO_3(s) \rightleftharpoons Pb^{2+}(aq) + CO_3^{2-}(aq) \qquad K_{sp} = 3.3 \times 10^{-14}$$

What is the concentration of CO_3^{2-} in a saturated lead carbonate solution?

Do your calculations here.

(2) Strontium sulfate, $SrSO_4$, is a slightly soluble salt that dissociates according to the following reaction:

$$SrSO_4(s) \rightleftharpoons Sr^{2+}(aq) + SO_4^{2-}(aq) \qquad K_{sp} = 3.8 \times 10^{-7}$$

What is the concentration of Sr^{2+} in a saturated $SrSO_4$ solution?

Do your calculations here.

9. (1) Tin (II) carbonate, $SnCO_3$, has a solubility of 5.73×10^{-3} g/L. Calculate the K_{sp}.

Do your calculations here.

(2) One form of zinc sulfide, ZnS (sphalerite), has a solubility of 6.4×10^{-4} g/L. Calculate the K_{sp}.

Do your calculations here.

(3) Calcium fluoride, CaF_2, has a solubility of 1.6×10^{-3} g per 100 mL of H_2O. Calculate the K_{sp} remembering that CaF_2 produces two F^- ions and one Ca^{2+} ion when it dissociates. The dissociation equation is

$$CaF_2(s) \rightleftharpoons Ca^{2+}(aq) + 2\, F^-(aq)$$

Do your calculations here.

10. Examine the formulas for the compounds shown. Make two lists ranking the acids from strongest to weakest and the salts from most soluble to least soluble. Then name all of the compounds.

HClO	$K_a = 3.5 \times 10^{-8}$
CaF_2	$K_{sp} = 3.9 \times 10^{-11}$
HCN	$K_a = 4.0 \times 10^{-10}$
$HC_2H_3O_2$	$K_a = 1.8 \times 10^{-5}$
$PbSO_4$	$K_{sp} = 1.3 \times 10^{-8}$
$Fe(OH)_3$	$K_{sp} = 6.0 \times 10^{-38}$

11. Calculate the H^+ ion concentration and the pH in a 0.1 M H_3PO_4 solution assuming the first ionization reaction is the primary reaction.

$$H_3PO_4 \leftrightarrows H^+ + H_2PO_4^- \qquad K_a = 7.5 \times 10^{-3}$$

Do your calculations here.

Challenge Problems

12. At 720 K, the K_{eq} for the following reaction has a value of 50.5.

    $$H_2(g) + I_2(g) \rightleftarrows 2\ HI(g)$$

    At the given temperature, what is the value for K_{eq} for the reverse reaction.

    $$2\ HI(g) \rightleftarrows H_2(g) + I_2(g)$$

    Calculate the equilibrium concentrations of all 3 species given a 1 L container and 0.0300 moles of HI gas at 720 K.

    Do your calculations here.

13. Calculate the pH of a buffer solution prepared by adding 0.10 mole of sodium acetate to 500 mL of a solution labeled 0.40 M acetic acid. K_a is 1.8×10^{-5}.

    Do your calculations here.

14. By using net ionic hydrolysis reactions, show why solutions of the following salts are either acidic or basic.

    (1) NH_4NO_3 – acidic
    (2) K_2CO_3 – basic
    (3) NaCN – basic
    (4) $Al_2(SO_4)_3$ – acidic

RECAP SECTION

Chapter 16 discusses two of the most interesting topics in chemistry — chemical equilibrium and chemical kinetics. The material in the text is especially important for you to read and discuss with your instructor. The exercises in the study guide are primarily of a problem-solving nature; the concepts have been covered in the text. Once equilibrium has been reached in a chemical reaction system, you can use Le Chatelier's Principle to predict what will happen to the system if reaction conditions are altered. Such predictions have important practical value in the chemical industry. In addition, you have learned to use the equilibrium constant expression to evaluate chemical reactions for ion concentrations and equilibrium constants.

Scientists in industry and research draw on the concepts of Chapter 16 constantly to help them solve practical chemical problems.

ANSWERS TO QUESTIONS AND SOLUTIONS TO PROBLEMS

1.
   (1) $2\ SO_2(g) + O_2(g) \rightleftarrows 2\ SO_3(g)$

   (2) $4\ NH_3(g) + 5\ O_2(g) \rightleftarrows 4\ NO(g) + 6\ H_2O(g)$

   (3) $N_2(g) + 3\ H_2(g) \rightleftarrows 2\ NH_3(g) + heat$

   (4) $3\ O_2(g) + heat \rightleftarrows 2\ O_3(g)$

   (5) $C(s) + CO_2(g) + heat \rightleftarrows 2\ CO(g)$

2.
   (1) a. Decreasing the amount of SO_2 will produces a shift to the left. b. Increasing the pressure will cause a shift to the right, which is the side of fewer molecules. c. A catalyst will have no effect on the equilibrium.

   (2) a. Decreasing the ozone concentration will shift the system to the right. b. So will increasing the amount of oxygen. c. Cooling the reaction will cause a shift to the left. d. Decreasing the pressure will cause a shift to the left.

   (3) a. Increasing the amount of carbon will not shift the equilibrium. b. Heating the mixture will shift the equilibrium to the right. c. Increasing the amount of CO will shift the reaction to the left. d. Increasing the amount of CO_2 will produce the opposite effect. e. Increasing the pressure will drive the reaction to the left.

3.
   (1) $K_{eq} = \dfrac{[SO_3]^2}{[SO_2]^2[O_2]}$

   (2) $K_{eq} = \dfrac{[O_3]^2}{[O_2]^3}$

   (3) $K_{eq} = \dfrac{[CO]^2}{[CO_2]}$

Carbon is in the solid state; therefore, it does not change concentration. It can be left out of the equilibrium expression.

(4) $K_{eq} = \dfrac{[H^+][NO_2^-]}{[HNO_2]}$

(5) $K_{eq} = \dfrac{[Al(NH_3)_2{}^+]}{[Al^{3+}][OH^-]}$

(6) $K_{eq} = \dfrac{[Al(OH)^{2+}]}{[Al^{3+}][OH^-]}$

(7) $K_{sp} = [Pb^{2+}][SO_4{}^{2-}]$

$PbSO_4$ is a solid; therefore, it is left out of the K_{sp} expression.

(8) $K_{eq} = \dfrac{[CO_2][H_2O]^2}{[CH_4][O_2]^2}$

(9) $K_{eq} = \dfrac{[H_2O]^2[Cl_2]^2}{[HCl]^4[O_2]}$

(10) $K_{eq} = \dfrac{[HClO][HCl]}{[Cl_2][H_2O]}$

4. $K_{eq} = 0.13$

The expression for the equilibrium constant, K_{eq}, would be:

$$K_{eq} = \dfrac{[SO_2]^2[O_2]}{[SO_3]^2}$$

If one-half of the SO_3 has decomposed at equilibrium, we can write our equilibrium conditions in tabular form.

Substance	Initial Conditions	Final Conditions
SO_3	0.5 M	0.25 M
SO_2	0 M	0.25 M
O_2	0 M	$\dfrac{0.25}{2}$ M

The equilibrium concentration of O_2 is one-half that of SO_2 according to the reaction. Substituting into the expression for K_{eq} we have

$$K_{eq} = \dfrac{[SO_2]^2[O_2]}{[SO_3]^2} = \dfrac{[0.25]^2[0.125]}{[0.25]^2} = 0.13$$

5. (1) 1.1×10^{-2} M or mol/L

The formula for nitrous acid is HNO_2, so the first step is to write out the ionization equation. Then formulate the equilibrium expression.

$$HNO_2 \rightleftarrows H^+ + NO_2^-$$

$$K_a = \frac{[H^+][NO_2^-]}{[HNO_2]}$$

$$4.5 \times 10^{-4} = \frac{[H^+][NO_2^-]}{[HNO_2]}$$

Now you are ready to determine your strategy for making substitutions in the equation.

One way to keep everything straight is to set up a small table such as you did for gas law problems.

Substance	Initial Conditions	Final Conditions
H^+	0 M	X
NO_2^-	0 M	X
HNO_2	0.25 M	0.25 – X

Each molecule of HNO_2 that ionizes produces 1 H^+ ion and 1 NO_2^- ion. Therefore, let their equilibrium concentration be "X" so that the amount of HNO_2 left at equilibrium will be 0.25 M – X. Substituting in the equation

$$4.5 \times 10^{-4} = \frac{[X][X]}{[0.25 - X]}$$

If X is small compared with 0.25 M, it is possible to simply the equation.

$$4.5 \times 10^{-4} = \frac{X^2}{0.25}$$

$$1.1 \times 10^{-4} = X^2$$

To find the value for X, take the square root of 1.1 and 10^{-4}. Refer to a math text or ask your instructor if you have difficulty with square roots.

Therefore,

$$X = 1.1 \times 10^{-2} \text{ M or mol/L} = [H^+]$$

X is small compared to 0.25 so you were justified in making your assumption.

(2) 1.0×10^{-2} M or mol/L

The formula for butyric acid is $CH_3(CH_2)_2COOH$, so the first step is to write out the ionization equation. After that formulate the equilibrium expression. The ionizable hydrogen in this case is shown on the right.

$$CH_3(CH_2)_2COOH \rightleftarrows CH_3(CH_2)_2COO^- + H^+$$

$$K_a = \frac{[CH_3(CH_2)_2COO^-][H^+]}{CH_3(CH_2)_2COOH}$$

Set up a table as before.

Substance	Initial Conditions	Final Conditions
H^+	0 M	X
$CH_3(CH_2)_2COO^-$	0 M	X
$CH_3(CH_2)_2COOH$	0.15 M	0.15 − X

Substituting in our equilibrium expression, you have the following equation

$$1.52 \times 10^{-5} = \frac{[X][X]}{[0.15 - X]}$$

Assuming the X is small compared with 0.15, you can simplify your equation.

$$1.52 \times 10^{-5} = \frac{X^2}{0.15}$$

$$1.01 \times 10^{-4} = X^2$$

Taking the square root of both sides without difficulty is possible since the roots are divisible by 2. Therefore,

$$X = 1.0 \times 10^{-2} \text{ M or mol/L}$$

X is small compared to 0.15 so you were justified in making your assumption.

6. (1) $H^+ = 1 \times 10^{-1}$ M or mol/L, $OH^- = 1 \times 10^{-13}$ M or mol/L. The problem states that the HBr is completely ionized. The equation would be

$$HBr \rightarrow H^+ + Br^-$$

This means that the H^+ concentration is the same as the HBr solution concentration or 0.1 M. In scientific notation, 0.1 M would be 1×10^{-1} M. To determine the OH^- concentration, we need to use the ion product constant of water relationship, which states that

$$[H^+][OH^-] = K_w = 1 \times 10^{-14}$$

We know what the concentration of H^+ ion is, so we can make the following substitution into the ion product constant equation

$$[1 \times 10^{-1}][OH^-] = 1 \times 10^{-14}$$

Therefore,
$$[OH^-] = \frac{1 \times 10^{-14}}{1 \times 10^{-1}} = 1 \times 10^{-13} \text{ M or mol/L}$$

Remember, when we are dividing exponents we subtract the denominator exponent from the numerator exponent.
$$-14 - (-1) = -14 + 1 = -13$$

(2) $H^+ = 1 \times 10^{-5}$ M or mol/L, $OH^- = 1 \times 10^{-9}$ M or mol/L. We can determine the H^+ ion concentration directly from the pH and then use the ion product constant of water to calculate the OH^- ion concentration.

The pH value is a whole number, which means that the H^+ concentration is 1 times a negative amount of 10. The value of the exponent is the same as the pH value. Therefore, if the pH is 5, the H^+ ion concentration is 1×10^{-5} M or mol/L. Substituting this value into the ion product constant expression

$$[H^+][OH^-] = K_w = 1 \times 10^{-14}$$
$$[1 \times 10^{-5}][OH^-] = 1 \times 10^{-14}$$
$$[OH^-] = \frac{1 \times 10^{-14}}{1 \times 10^{-5}} = 1 \times 10^{-9} \text{ M or mol/L}$$

(3) pH = 10

There are at least two ways to solve this problem. By one method, we first need to find out what the H^+ concentration is. We can use the ion product constant expression to do this.

$$[H^+][OH^-] = K_w = 1 \times 10^{-14}$$
$$[H^+][1 \times 10^{-14}] = 1 \times 10^{-14}$$
$$[H^+] = \frac{1 \times 10^{-14}}{1 \times 10^{-4}} = 1 \times 10^{-10} \text{ M or mol/L}$$

We can easily convert this value to pH since the value in front of the exponential value is 1 (one). Therefore, the pH is 10. Another way to solve the problem is to use the relationship

$$pH + pOH = 14$$

Converting OH^- concentration into pOH, we have

$$1 \times 10^{-4} = \text{pOH of 4}$$

Then, we can substitute the pOH value

$$pH + 4 = 14$$
$$pH = 14 - 4 = 10$$

(4) a.   pOH = 10

   $[H^+] = 1 \times 10^{-4}$ M

   $[OH^-] = 1 \times 10^{-10}$ M

   b. $[OH^-] = 1 \times 10^{-8}$ M

   pH = 6

   $[H^+] = 1 \times 10^{-6}$ M

   c.   pH = 11

   pOH = 3

   $[OH^-] = 1 \times 10^{-3}$ M

(5) a. buffer   b. acetate   c. acetic acid
     d. $HC_2H_3O_2$ e. $H_2O + C_2H_3O_2^-$

(6) a and e, b and d, c and f

7. When H^+ (acid) is added to the phosphate buffer system, the additional H^+ reacts with HPO_4^{2-} to produce undissociated $H_2PO_4^-$, which is a weak acid.

$$H^+ + HPO_4^{2-} \rightleftharpoons H_2PO_4^-$$

The additional HPO_4^{2-} necessary is present because of the complete ionization of Na_2HPO_4.

On the other hand when OH^- (base) is added, it reacts with undissociated $H_2PO_4^-$.

$$H_2PO_4^- = OH^- \rightleftharpoons H_2O + HPO_4^{2-}$$

Since H_2O dissociates to only a small extent, the excess OH^- is used up by reacting with the weak acid $H_2PO_4^-$.

8. (1) $CO_3^{2-} = 1.8 \times 10^{-7}$ M or mol/L
   First, write the equilibrium expression with the appropriate value for the solubility product

   $$PbCO_3 \rightleftharpoons Pb^{2+} + CO_2^{2-}$$

   $$K_{sp} = 3.3 \times 10^{-14} = [Pb^{2+}][CO_3^{2-}]$$

   Since $PbCO_3$ is a solid whose concentration is unchanging in the equilibrium system you can cancel the term $[PbCO_3(s)]$ out of the equation. Since the concentration of Pb^{2+} is equal to the concentration of CO_3^{2-} at equilbrium, you can substitute "X" into the K_{sp} equation for both ion concentrations.

   $3.3 \times 10^{-14} = [X][X]$

   $3.3 \times 10^{-14} = X^2$

   $1.8 \times 10^{-7} = X$

   Therefore,

   $[Pb^{2+}] = [CO_3^{2-}] = 1.8 \times 10^{-7}$ M

(2) $Sr^{2+} = 6.2 \times 10^{-4}$ M
The equation is
$$K_{sp} = 3.8 \times 10^{-7} = [Sr^{2+}][SO_4^{2-}]$$
Since
$$[Sr^{2+}] = [SO_4^{2-}] = X$$
Therefore:
$$3.8 \times 10^{-7} = [X][X]$$
$$3.8 \times 10^{-7} = X^2$$
$$6.2 \times 10^{-4} = X$$
Therefore:
$$[Sr^{2+}] = [SO_4^{2-}] = 6.2 \times 10^{-4} \text{ M}$$

9. (1) $K_{sp} = 1.03 \times 10^{-9}$

First calculate the molar mass of $SnCO_3$, using a list of atomic masses.

$$\text{molar mass of } SnCO_3 = (118.7 \text{ g}) + (12.01 \text{ g}) + (3 \times 16.00 \text{ g}) = 178.7 \frac{g}{mol}$$

Next, determine the molarity of the saturated solution of $SnCO_3$, then substitute the values in the solubility product equation.

$$\text{Molarity} = \frac{g/L}{g/mol} = \frac{mol}{L}$$

Substituting in the above equation

$$M = \frac{5.73 \times 10^{-3} \text{ g/L}}{178.7 \text{ g/mol}} = 3.21 \times 10^{-5} \text{ M}$$

$$K_{sp} = [Sn^{2+}][CO_3^{2-}], [Sn^{2+}] = [CO_3^{2-}] = 3.21 \times 10^{-5} \text{ M}$$
$$= [3.21 \times 10^{-5}][3.21 \times 10^{-5}]$$
$$= 10.3 \times 10^{-10}$$
$$K_{sp} = 1.03 \times 10^{-9}$$

(2) $K_{sp} = 4.4 \times 10^{-11}$

$$\text{molar mass of } ZnS = 97.44 \frac{g}{mol}$$

$$M = \frac{g/L}{g/mol} = \frac{6.4 \times 10^{-4} \text{ g/L}}{97.44 \text{ g/mol}} = 6.6 \times 10^{-6} \text{ M}$$

$$K_{sp} = [Zn^{2+}][S^{2-}], [Zn^{2+}] = [S^{2-}] = 6.6 \times 10^{-6} \text{ M}$$
$$= [6.6 \times 10^{-6}][6.6 \times 10^{-6}]$$
$$= 44 \times 10^{-12}$$
$$K_{sp} = 4.4 \times 10^{-11}$$

(3) $K_{sp} = 3.2 \times 10^{-11}$

First calculate the molar mass of CaF_2, using a list of atomic masses.

$$\text{molar mass of } CaF_2 = 40.08 \text{ g} + (2 \times 19.00 \text{ g}) = 78.08 \frac{\text{g}}{\text{mol}}$$

A solubility of 1.6×10^{-3} g per 100 mL of H_2O is the same as 1.6×10^{-2} g per L. Using this number, we can determine the molarity of a saturated solution of CaF_2, assuming the volume of water to be identical with the volume of the solution.

$$\text{Molarity} = \frac{\text{moles}}{L}$$

Substituting in the above equation.

$$M = \frac{1.6 \times 10^{-2} \text{ g/L}}{78.08 \text{ g/mol}} = 2.0 \times 10^{-4} \text{ M}$$

$$CaF_2(s) \rightleftharpoons Ca^{2+}(aq) + 2 \text{ F}^-(aq)$$

$$K_{sp} = [Ca^{2+}][F^-]^2$$

$$= [2.0 \times 10^{-4}][2(2.0 \times 10^{-4})]^2$$

Remember, there are two F^- ions for each Ca^{2+} ion.

$$K_{sp} = [2.0 \times 10^{-4}][4.0 \times 10^{-4}]^2$$

$$= 32 \times 10^{-12}$$

$$= 3.2 \times 10^{-11}$$

10. Acids $HC_2H_3O_2$, acetic acid, is strongest
HClO, hypochlorous acid, is next
HCN, hydrocyanic, is weakest

Salts CaF_2, calcium fluoride, is most soluble
$PbSO_4$, lead(II) sulfate, is next
$Fe(OH)_3$, iron(III) hydroxide, is least soluble

11. H^+ concentration equals 2.74×10^{-2} M, pH is approximately 1.56. Establish a table as before.

Substance	Initial Conditions	Final Conditions
H_3PO_4	0.100 M	0.100 − X
$H_2PO_4^-$	0 M	X
H^+	0 M	X

Substitute the final values into the expression for K_a and assume X has a small value compared to the initial concentration of H_3PO_4.

$$K_a = 7.5 \times 10^{-3} = \frac{[H^+][H_2PO_4^-]}{[H_3PO_4]} = \frac{[X][X]}{[0.1 - X]}$$

$$7.5 \times 10^{-3} = \frac{X^2}{0.1}$$

$$[H^+] = X = \sqrt{7.5 \times 10^{-4}} = 2.74 \times 10^{-2}$$
$$pH = -\log[H^+] = 1.56$$

12. $K_{eq} = 1.98 \times 10^{-2}$, $[H_2] = [I_2] = 4.22 \times 10^{-3}$ M, $[HI] = 2.16 \times 10^{-2}$ M. Quadratic formula yields $[H_2] = [I_2] = 5.88 \times 10^{-3}$ M, $[HI] = 1.83 \times 10^{-2}$ M. The K_{eq} for the reverse reaction is the reciprocal of the K_{eq} for the forward reaction.

$$K_{eq}(\text{reverse}) = \frac{1}{K_{eq}(\text{forward})} = \frac{1}{50.5} = 1.98 \times 10^{-2}$$

The initial concentration of HI is 0.0300 M. Set up a table of values as before

Substance	Initial Conditions	Final Conditions
HI	0.0300 M	$0.0300 - X$
H_2	0 M	$\frac{X}{2}$
I_2	0 M	$\frac{X}{2}$

One mole of HI decomposes to produce 0.5 mole each of H_2 and I_2.

$$K_{eq} = \frac{[H_2][I_2]}{[HI]^2} = \frac{\left[\frac{X}{2}\right]\left[\frac{X}{2}\right]}{[0.0300 - X]^2} = \frac{\left[\frac{X^2}{4}\right]}{9.00 \times 10^{-4}} = 1.98 \times 10^{-2}$$

Assume X is small compared to 0.0300 M.

Simplifying

$$(1.98 \times 10^{-2})(9.00 \times 10^{-4})(4) = X^2$$
$$X^2 = 7.13 \times 10^{-5}$$
$$X = 8.44 \times 10^{-3}$$

If $X = 8.44 \times 10^{-3}$, then the concentrations of H_2 and I_2 are X/2 or 4.22×10^{-3} M. The concentration of HI is 0.0300 M − 0.00844 M or 0.0216 M.

The value for X is not really small compared to 0.0300 M. Using the quadratic formula, a value for X of 1.17×10^{-2} is obtained rather than 8.44×10^{-3}. In this case the concentrations of H_2 and I_2 would be calculated to be 5.85×10^{-3} M and the concentration of HI would be 0.0183 M.

13. pH = 4.4

The equation is $HC_2H_3O_2 \rightleftharpoons H^+ + C_2H_3O_2^-$

$$K_a = \frac{[H^+][C_2H_3O_2^-]}{[HC_2H_3O_2]} = 1.8 \times 10^{-5}$$

0.1 mole sodium acetate furnishes 0.1 mole acetate ion.

$$NaC_2H_3O_2 \rightarrow Na^+ + C_2H_3O_2^-$$

$$\frac{0.10 \text{ mol}}{500 \text{ mL}} = \frac{0.20 \text{ mol}}{1000 \text{ mL}} = \frac{0.20 \text{ mol}}{\text{L}}$$

Substituting into the equation we have

$$\frac{[\text{H}^+][0.20]}{[0.40]} = 1.8 \times 10^{-5}$$

$$\text{H}^+ = \frac{(1.8 \times 10^{-5})(0.40)}{0.20} = 3.6 \times 10^{-5} \text{ M}$$

$$\text{pH} = -\log[\text{H}^+] = 4.4$$

The pH can also be calculated from an equation named the Henderson-Hasselbach equation

$$\text{pH} = \text{pKa} + \log \frac{[\text{salt}]}{[\text{acid}]}$$

The answer will be the same either way.

14. (1) $\text{NH}_4^+ + \text{H}_2\text{O} \rightleftharpoons \text{NH}_4\text{OH} + \text{H}^+$

   (2) $\text{CO}_3^{2-} + 2\,\text{H}_2\text{O} \rightleftharpoons \text{H}_2\text{CO}_3 + 2\,\text{OH}^-$

   (3) $\text{CN}^- + \text{H}_2\text{O} \rightleftharpoons \text{HCN} + \text{OH}^-$

   (4) $2\,\text{Al}^{3+} + 6\,\text{H}_2\text{O} \rightleftharpoons 2\,\text{Al(OH)}_3 + 6\,\text{H}^+$

WORD SEARCH 2

In the given matrix, find the terms that match the following definitions. Terms may be horizontal, vertical, or on the diagonal. They may also be written forward or backward. Answers are found at the end of chapter 20.

1. A principle that states what happens to an equilibrium system when the conditions are altered.
2. A solution that resists changes in pH.
3. An experimental technique for measuring the volume of one reagent required to react with a measured amount of another reagent.
4. The formation of ions.
5. A dynamic state where two or more opposing processes are taking place at the same time and at the same rate.
6. A homogeneous mixture of two or more substances.
7. A substance whose aqueous solution conducts electricity.
8. Capable of mixing and forming a solution.
9. The H_3O^+ ion.
10. A solution containing 1 mole of solute per liter of solution is 1.0 _____ .
11. A solution containing dissolved solute in equilibrium with undissolved solute.
12. Behaving chemically as either an acid or base.
13. The substance present to the largest extent in a solution.
14. The number of equivalent weights of solute per liter of solution.
15. Incapable of mixing.
16. The substance that is dissolved in a solvent to form a solution.
17. Solution containing a relatively small amount of solute.

A	E	G	Y	T	F	E	H	Y	K	B	D	I	S	P	D	N	J
G	P	I	S	O	L	U	T	E	O	Q	F	Y	D	I	L	K	Z
X	T	D	N	O	A	I	S	Y	C	I	E	V	L	A	H	P	O
C	K	E	B	G	L	O	M	R	L	X	R	U	N	F	O	J	Z
U	M	T	V	A	E	V	Z	M	A	O	T	E	Q	B	Q	Y	M
S	F	A	M	H	C	R	E	D	I	E	R	E	F	F	U	B	Q
D	N	R	U	S	H	E	R	N	C	S	Q	T	U	W	M	J	P
C	O	U	I	J	A	L	B	P	T	L	C	R	C	G	H	Z	W
N	I	T	N	U	T	B	I	N	E	H	A	I	V	E	E	Y	I
X	T	A	O	W	E	I	C	T	O	L	R	T	B	A	L	O	Q
K	A	S	R	D	L	C	V	M	O	E	W	M	Z	L	V	E	N
V	R	G	D	L	I	S	E	M	T	B	L	N	E	C	E	B	L
O	T	F	Y	X	E	I	I	O	N	I	Z	A	T	I	O	N	R
Y	I	G	H	W	R	M	H	Z	A	X	U	F	A	K	T	T	F
I	T	S	V	A	X	P	U	N	O	I	T	U	L	O	S	G	J
J	P	L	Q	Z	M	U	I	R	B	I	L	I	U	Q	E	E	T
M	C	X	S	A	H	T	A	U	D	F	N	Y	P	R	B	R	W

CHAPTER SEVENTEEN

Oxidation-Reduction

SELF-EVALUATION SECTION

1. The following are selected rules for determining oxidation numbers of elements in compounds or ions.

   Hydrogen is generally +1.
   Oxygen is generally −2.

   The algebraic sum of the oxidation numbers for all the atoms in a compound is equal to zero.

   The algebraic sum of the oxidation numbers for all the atoms in a polyatomic ion is equal to the charge of the ion.

   Determine the oxidation number for each indicated element in the following compounds or ions:

   (1) S in H_2SO_3 _____   (9) Mn in $MnO(OH)_2$ _____
   (2) N in HNO_3 _____   (10) C in H_2CO_3 _____
   (3) Cr in $Cr_2O_7^{2-}$ _____   (11) C in $C_2O_4^{2-}$ _____
   (4) Cl in ClO_4^- _____   (12) As in AsO_4^{3-} _____
   (5) Zn in ZnO_2^{2-} _____   (13) P in P_2O_5 _____
   (6) Mn in MnO_4^- _____   (14) N in N_2O _____
   (7) C in CH_4 _____   (15) N in NO_2^- _____
   (8) C in $C_6H_{12}O_6$ _____   (16) Br in BrO_3^- _____

2. In the following reactions, identify the element that is *reduced*, the element that is *oxidized*, the *reducing agent*, and the *oxidizing agent*.

Oxidation Information

Elements in the free state have an oxidation number of 0.
Hydrogen is usually +1.
Oxygen is usually −2.
Metallic elements in ionic compounds have a positive charge.
Group IA metals are always +1.

(1) $Cr_2O_3 + 3\ H_2(g) \xrightarrow{\Delta} 2\ Cr(s) + 3\ H_2O(l)$

(2) $2\ H_2(g) + O_2(g) \xrightarrow{\Delta} 2\ H_2O(l)$

(3) $2\ HNO_2(aq) + 2\ HI(aq) \rightarrow I_2(s) + 2\ NO(g) + 2\ H_2O(l)$

(4) $2\ Na(s) + H_2O(l) \rightarrow 2\ NaOH(aq) + H_2(g)$

(5) $5\ NaBr + NaBrO_3 + 3\ H_2SO_4 \rightarrow 3\ Br_2 + 3\ Na_2SO_4 + 3\ H_2O$

Write your answers here.

3. Using the oxidation number information given for question 1, balance the following un-ionized and ionic equations using the electron gain-and-loss technique. Place correct coefficients in front of each formula.

(1) $HNO_3(aq) + HI(aq) \rightarrow NO(g) + I_2(s) + H_2O(l)$

(2) $Mn^{4+}(aq) + Cl^-(aq) \rightarrow Cl_2(g) + Mn^{2+}(aq)$

(3) $ClO_3^-(aq) + SnO_2^{2-}(aq) \rightarrow Cl^-(aq) + SnO_3^{2-}(aq)$

(4) $NH_3(g) + O_2(g) \rightarrow NO(g) + H_2O(g)$

(5) $H^+(aq) + Br^-(aq) + SO_4^{2-}(aq) \rightarrow Br_2(l) + SO_2(g) + H_2O(l)$

Write your answers here.

4. Balance the following redox equations by the ion-electron method. For acidic conditions use H^+ and H_2O to balance H and O. For basic conditions use OH^- and H_2O.

   (1) $Zn(s) + NO_3^-(aq) \rightarrow Zn^{2+}(aq) + NH_4^+(aq)$ (acidic)
   (2) $ClO^- + I^- \rightarrow IO_3^- + Cl^-$ (basic)

   Write your answers here.

5. Using the partial list of the activity series of metals, answer the following questions. For any reaction that proceeds write a balanced equation.

   $Mg \rightarrow Mg^{2+} + 2e^-$
   $Al \rightarrow Al^{3+} + 3e^-$
   $Zn \rightarrow Zn^{2+} + 2e^-$
   $Fe \rightarrow Fe^{2+} + 2e^-$
   $Ni \rightarrow Ni^{2+} + 2e^-$
   $Sn \rightarrow Sn^{2+} + 2e^-$
   $Pb \rightarrow Pb^{2+} + 2e^-$
   $H_2 \rightarrow 2 H^+ + 2e^-$
   $Cu \rightarrow Cu^{2+} + 2e^-$
   $Ag \rightarrow Ag^+ + e^-$

   (1) Will a chemical reaction occur when a $Pb(NO_3)_2$ solution ($Pb^{2+} + 2 NO_3^-$) is placed in a *copper* container?
   (2) Will a chemical reaction occur when a $ZnCl_2$ solution ($Zn^{2+} + 2 Cl_2^-$) is placed in a *magnesium* container?
   (3) Will a chemical reaction occur when a HCl solution ($H^+ + Cl^-$) is placed in a *tin* can?
   (4) Will a chemical reaction occur when an $AgNO_3$ solution ($Ag^+ + NO_3^-$) is placed in an *aluminum* container?

Do your equations here.

6. Aluminum metal is produced almost entirely by the electrolysis of Al_2O_3. The simplified reactions are

    $Al^{3+} + 3e^- \rightarrow Al$

    $O^{2-}(aq) + C(s) \rightarrow CO(g) + 2e^-$

    In the sketch below, indicate the direction of ion migration and balance the overall equation.

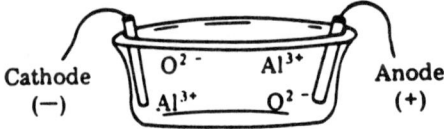

Cathode (−)   Anode (+)

Write your answers here.

7. (1) Write the anode and cathode reactions for the electrolysis of molten $NiBr_2$ with inert electrodes. Then write the balanced overall reaction.

    Write your answers here.

(2) Write anode, cathode, and the overall reaction for the electrolysis of molten LiCl with inert electrodes.

Write your answers here.

8. Identify which half-reaction of each pair occurs at the anode and which half-reaction occurs at the cathode.

   (1) $6\ Fe^{2+} \rightarrow 6\ Fe^{3+} + 6e^-$

   $Cr_2O_7^{2-} + 14\ H^+ + 6e^- \rightarrow 2\ Cr^{3+} + 7\ H_2O$

   (2) $2\ H^+ + H_2O_2 + 2e^- \rightarrow 2\ H_2O$

   $2\ I^- \rightarrow I_2 + 2e^-$

   (3) $Fe^0 \rightarrow Fe^{2+} + 2e^-$

   $Ni_2O_3 + 3\ H_2O + 2e^- \rightarrow 2\ Ni(OH)_2 + 2\ OH^-$

Write your answers here.

9. Fill in the blank space with a correct term.

   The type of cell that uses chemical reactions to produce electrical energy is called a (1) _____ cell. the other type of cell, (2) _____, uses electrical energy to produce a chemical change. A dry cell flashlight battery is an example of a(an) (3) _____ .

10. The following reactions are involved in the operation of the lead storage battery. Which is the oxidation reaction and which is the reduction reaction? Give the overall balanced cell reaction.

   (1) $Pb \rightarrow Pb^{2+} + 2e^-$

   (2) $PbO_2 + 4\,H^+ + 2e^- \rightarrow Pb^{2+} + 2\,H_2O$

Challenge Problems

11. The reaction used to measure the biochemical oxygen demand of a water sample involves a 0.025 N solution of sodium thiosulfate according to the following reaction.

    $$2\,Na_2S_2O_3 \cdot 5\,H_2O + I_2 \rightarrow Na_2S_4O_6 + 2\,NaI + 10\,H_2O$$

    Determine the mass of sodium thiosulfate pentahydrate necessary to prepare 1 L of a 0.025 N solution for the above reaction.

    Do your calculations here.

12. Balance the following oxidation-reduction equations. The reaction conditions are either acidic or basic, and H^+ and OH^- plus H_2O should be used accordingly to balance the reactions.

    (1) $MnO_4^- + VO^{2+} \rightarrow VO_2^+ + Mn^{2+}$ (acid)

    (2) $P_4 \rightarrow PH_3 + H_2PO_2^-$ (basic solution)

    (3) $MnO_2 + SO_3^{2-} \rightarrow SO_4^{2-} + Mn(OH)_2$ (basic solution)

    (4) $MnO_4^-\,(aq) + H_2O_2(l) \rightarrow Mn^{2+}(aq) + O_2(g)$ (acid)

    (5) $Mn^{2+} + HBiO_3 \rightarrow Bi^{3+} + MnO_4^-$ (acid)

    (6) $Zn(s) \rightarrow Zn(OH)_4^{2-}(aq) + H_2(g)$ (basic solution)

13. Determine the oxidation number of the indicated element in each of the following compounds.

   (1) Cr in K_2CrO_4 _____
   (2) S in $Na_2S_2O_3$ _____
   (3) Ti in $NaTi_3O_7$ _____
   (4) Pt in $H_2PtCl_6 \cdot 6H_2O$ _____
   (5) C in $H_2C_2O_4$ _____
   (6) Co in $Na_3Co(NO_2)_6$ _____
   (7) S in $(NH_4)_2S_2O_8$ _____
   (8) B in CaB_4O_7 _____
   (9) Mo in $(NH_4)_2MoO_4$ _____
   (10) As in $HAsO_3$ _____

RECAP SECTION

The major thrust of Chapter 17 is to give each learner experience at working with chemical equations and manipulating numbers. If you answered the study guide questions without difficulty, you have mastered a valuable chemical tool. You are able to take equations apart, analyze them, and make them work for you. You should realize by now that you have all the skills needed to take a word equation and transform it into a balanced tool to be used for chemical description and calculations. You can handle ions, molecules, or gases in equations and calculations. If you have had difficulty, you should go back over the examples in the text very carefully and then come back to the study guide and the review problems at the end of the chapter.

ANSWERS TO QUESTIONS AND SOLUTIONS TO PROBLEMS

1. (1) +4 (2) +5 (3) +6 (4) +7 (5) +2 (6) +7 (7) −4 (8) 0
   (9) +4 (10) +4 (11) +3 (12) +5 (13) +5 (14) +1 (15) +3 (16) +5

2. (1) Hydrogen gas is oxidized and is the reducing agent. Chromium is reduced and is the oxidizing agent.

$$Cr_2O_3(s) + 3\ H_2(g) \xrightarrow{\Delta} 2\ Cr(s) + 3\ H_2O(l)$$

First, assign oxidation numbers. O = −2 and H = +1. Hydrogen gas = 0. Chromium metal = 0.

$$Cr_2O_3 + 3\ H_2 \rightarrow 2\ Cr + 3\ H_2O$$
$${+3\ -2} {0} {0} {+1\ -2}$$

What is Cr in Cr_2O_3? If $O = -2$, then each Cr will have to be +3 in order to balance the total amount of negative oxidation value of 3 oxygen. Which element has been oxidized (lost electrons)? Hydrogen has changed from 0 to +1. so it has been oxidized. Cr has changed from +3 to 0, so it has been reduced.

(2) Hydrogen gas oxidized and is the reducing agent. Oxygen gas is reduced and is the oxidizing agent.

$$2\ H_2(g) + O_2(g) \xrightarrow{\Delta} 2\ H_2O(g)$$
$${0}{0}\phantom{O_2(g) \xrightarrow{\Delta} 2\ }{+1\ -2}$$

Each hydrogen atom has lost an electron, and each oxygen atom has gained two electrons.

(3) Iodine is oxidized and is the reducing agent. Nitrogen is reduced and is the oxidizing agent.

$$2\ HNO_2(aq) + 2\ HI(aq) \rightarrow I_2(s) + 2\ NO(g) + 2\ H_2O(l)$$
$${+1\ +3\ -2} {+1\ -1} {0} {+2\ -2} {+1\ -2}$$

Iodine changes from −1 to 0, which is a loss of one electron (oxidation). Nitrogen has gained one electron in changing from +3 to +2.

(4) Sodium is oxidized and hydrogen is reduced. Sodium is the reducing agent, and hydrogen is the oxidizing agent.

$$2\ Na + H_2O \rightarrow 2\ NaOH + H_2$$
$${0} {+1\ -2} {+1\ -2\ +1} {0}$$

Sodium changes from 0 to +1, an oxidation process. Na is the reducing agent. Hydrogen changes from +1 to 0 for one of the products, hydrogen gas. Notice that hydrogen is also present in NaOH at the same oxidation state as in H_2O. Hydrogen is reduced and is the oxidizing agent.

(5) Bromine is both oxidized and reduced.

$$5\ NaBr + NaBrO_3 + 3\ H_2SO_4 \rightarrow 3\ Br_2 + 3\ Na_2SO_4 + 3\ H_2O$$
$${+1\ -1} {+1\ +5\ -2} {+1\ +6\ -2} {0} {+1\ +6\ -2}$$

The only atoms changing oxidation number are Br. Five Br atoms change from −1 to 0. This is oxidation, a loss of electrons. These Br atoms are the reducing agents, The other Br atom changes from +5 to 0, which is a gain of electrons, or reduction. This Br atom is the oxidizing agent.

3. (1) $2\ HNO_2(aq) + 2\ HI(aq) \rightarrow 2\ NO(g) + I_2(s) + 2\ H_2O(l)$

Assign oxidation numbers and write each half-reaction for oxidation and reduction.

$$HNO_2 + HI \rightarrow NO + I_2 + H_2O$$
$${+1\ +3\ -2} {+1\ -1} {+2\ -2} {0} {+1\ -2}$$

Oxidation $I^- \to I_2$

balance $2\,I^- \to I_2 + 2e^-$ (I^- loses 1 electron)

Reduction $N^{3+} \to N^{2+}$

balance $N^{3+} + 1e^- \to N^{2+}$ (N^{3+} gains 1 electron)

Multiply the reduction equation by 2 to balance the electron gain and loss.

$2\,I^- \to I_2 + 2e^-$

$2\,N^{3+} + 2e^- \to 2\,N^{2+}$

Place each half-reaction along with the coefficient back in the original equation.

$2\,HNO_2(aq) + 2\,HI(aq) \to 2\,NO(g) + I_2(s) + H_2O(l)$

Balance the remaining H and O atoms.

$2\,HNO_2(aq) + 2\,HI(aq) \to 2\,NO(g) + I_2(s) + 2\,H_2O(l)$

(2) $Mn^{4+}(aq) + 2\,Cl^-(aq) \to Cl_2(g) + Mn^{2+}(aq)$

This equation should be fairly simple to balance.

$Mn^{4+}(aq) + Cl^-(aq) \to Cl_2(g) + Mn^{2+}(aq)$

Cl^- is oxidized to free elemental chlorine, and Mn^{4+} is reduced to Mn^{2+}.

Oxidation $2\,Cl^-(aq) \to Cl_2(g) + 2e^-$ (1 electron lost per atom, 2 electrons lost per molecule)

Reduction $Mn^{4+}(aq) + 2e^- \to Mn^{2+}(aq)$ (2 electrons gained per atom)

$Mn^{4+}(aq) + 2\,Cl^-(aq) \to Cl_2(g) + Mn^{2+}(aq)$

The equation is now balanced. Check the electrical charges for the last detail.

$(4+) + (2-) = 2+$
  left side   right side

The equation must have the same electrical charge on both sides as well as the same number of atoms of each element.

(3) $ClO_3^-(aq) + 3\,SnO_2^{2-}(aq) \to Cl^-(aq) + 3\,SnO_3^{2-}(aq)$

Assign oxidation numbers and write each half-reaction for oxidation and reduction.

$ClO_3^-(aq) + SnO_2^{2-}(aq) \to Cl^-(aq) + SnO_3^{2-}(aq)$
  +5-2   +2-2   -1   +4-2

Oxidation $Sn^{2+}(aq) \to Sn^{4+}(aq) + 2e^-$ ($2e^-$ loss per atom)

Reduction $Cl^{5+}(aq) + 6e^- \to Cl^-(aq)$ ($6e^-$ gain per atom)

Balance the electron gain and loss first.

$3\,Sn^{2+}(aq) \to 3\,Sn^{4+}(aq) + 6e^-$

$Cl^{5+}(aq) + 6e^- \to Cl^-(aq)$

Place each coefficient back in the original equation.

$ClO_3(aq) + 3\,SnO_2^{2-}(aq) \to Cl^-(aq) + 3\,SnO_3^{2-}(aq)$

Check whether the atoms of oxygen balance. Do the electrical charges balance also?

$$(-1) + (6-) = (-1) + (6-)$$
$$\text{left side} \qquad \text{right side}$$

(4) $4\,NH_3(g) + 5\,O_2(g) \rightarrow 4\,NO(g) + 6\,H_2O(g)$

Assign oxidation numbers and write each half-reaction for oxidation and reduction.

$$NH_3(g) + O_2(g) \rightarrow NO(g) + H_2O(g)$$
$$\;-3+1\qquad\;\;0\qquad\;\;+2-2\quad\;+1-2$$

This equation has a small wrinkle to it. Oxygen is reduced from 0 to −2 and shows up in two molecules on the product side

Oxidation $\quad N^{3-} \rightarrow N^{2+} + 5e^-$

Reduction $\quad O_2^{\,0} + 4e^- \rightarrow 2\,O^{2-}\qquad$ (from NO and H$_2$O)

To balance a gain of 4e$^-$ and a loss of 5e$^-$, we must use 20 as the lowest common denominator.

$$4\,N^{3-} \rightarrow 4\,N^{2+} + 20e^-$$
$$5\,O_2 + 20e^- \rightarrow 10\,O^{2-}$$

Put coefficients back in the equation. Remember that we can only have four NO molecules and that the rest of the oxygen on the product side is in water.

$$4\,NH_3(g) + 5\,O_2(g) \rightarrow 4\,NO(g) + H_2O(g)$$

To balance out the hydrogen, place a 6 in front of the water. The oxygen is now also balanced.

$$4\,NH_3(g) + 5\,O_2(g) \rightarrow 4\,NO(g) + 6\,H_2O(g)$$

(5) $4\,H^+(aq) + 2\,Br^-(aq) + SO_4^{2-}(aq) \rightarrow Br_2(l) + SO_2(g) + 2\,H_2O(l)$

Assign oxidation numbers and write each half-reaction for oxidation and reduction.

$$H^+(aq) + Br^-(aq) + SO_4^{2-}(aq) \rightarrow Br_2(l) + SO_2(g) + H_2O(l)$$
$$\;+1\qquad\;-1\qquad\;+6-2\qquad\qquad\;0\qquad\;+4-2\quad\;+2-2$$

Oxidation $\;2\,Br^- \rightarrow Br_2 + 2e^-$

Reduction $\;S^{6+} + 2e^- \rightarrow S^{4+}$

Electron gain and loss is balanced. Insert a 2 in front of Br$^-$.

$$H^+(aq) + 2\,Br^-(aq) + SO_4^{2-}(aq) \rightarrow Br_2(l) + SO_2(g) + H_2O(l)$$

Check for balance of atoms. H and O are still out of balance. We need four H$^+$ on the left side and two H$_2$O molecules on the right.

$$4\,H^+(aq) + 2\,Br^-(aq) + SO_4^{2-}(aq) \rightarrow Br_2(l) + SO_2(g) + 2\,H_2O(l)$$

Do a check on electrical charges.

$$(4+) + (2-) + (2-) = 0$$
$$\text{left side}\qquad\text{right side}$$

4. (1) $4 \text{ Zn} + \text{NO}_3^- + 10 \text{ H}^+ \rightarrow 4 \text{ Zn}^{2+} + \text{NH}_4^+ + 3 \text{ H}_2\text{O}$

To begin, write the two half-reactions containing the elements being oxidized and reduced.

Oxidation $\quad \text{Zn} \rightarrow \text{Zn}^{2+}$

Reduction $\quad \text{NO}_3^- \rightarrow \text{NH}_4^+$

The Zn and N are balanced on each side, so now balance the H and O. Remember acid conditions.

$$10 \text{ H}^+ + \text{NO}_3^- \rightarrow \text{NH}_4^+ + 3 \text{ H}_2\text{O}$$

Now balance the electrons in the NO_3^- half-reaction.

$$8e^- + 10 \text{ H}^+ + \text{NO}_3^- \rightarrow \text{NH}_4^+ + 3 \text{ H}_2\text{O}$$

We now have $8e^-$ available for the oxidation half-reaction of Zn solid. This means we can use four atoms of Zn for each NO_3^-.

$$4 \text{ Zn} \rightarrow \text{Zn}^{2+} + 8e^-$$

Adding the two half-reactions together we have

$$4 \text{ Zn} + \text{NO}_3^- + 10 \text{ H}^+ \rightarrow 4 \text{ Zn}^{2+} + \text{NH}_4^+ + 3 \text{ H}_2\text{O}$$

(2) $3 \text{ ClO}^- + \text{I}^- \rightarrow 3 \text{ Cl}^- + \text{IO}_3^-$

Write the two half-reactions

Oxidation $\quad \text{I}^- \rightarrow \text{IO}_3^-$

Reduction $\quad \text{ClO}^- \rightarrow \text{Cl}^-$

The I and Cl are balanced. Now proceed with the H and O using basic conditions.

To balance O and H in the oxidation equation, add $3 \text{ H}_2\text{O}$ to the left and 6H^+ to the right side of the equation.

$$\text{I}^- + 3 \text{ H}_2\text{O} \rightarrow \text{IO}_3^- + 6 \text{ H}^+$$

Add 6 OH^- to each side

$$6 \text{ OH}^- + \text{I}^- + 3 \text{ H}_2\text{O} \rightarrow \text{IO}_3^- + 6 \text{ H}^+ + 6 \text{ OH}^-$$

Combine $\text{H}^+ + \text{OH}^- \rightarrow \text{H}_2\text{O}$

$$6 \text{ OH}^- + \text{I}^- + 3 \text{ H}_2\text{O} \rightarrow \text{IO}_3^- + 6 \text{ H}_2\text{O}$$

Rewrite canceling H_2O on each side

$$6 \text{ OH}^- + \text{I}^- \rightarrow \text{IO}_3^- + 3 \text{ H}_2\text{O}$$

To balance O and H in the reduction equation, add $1 \text{ H}_2\text{O}$ to the right side of the equation and 2 H^+ to the left side.

$$2 \text{ H}^+ + \text{ClO}^- \rightarrow \text{Cl}^- + \text{H}_2\text{O}$$

Add 2 OH^- to each side.

$$2 \text{ OH}^- + 2 \text{ H}^+ + \text{ClO}^- \rightarrow \text{Cl}^- + \text{H}_2\text{O} + 2 \text{ OH}^-$$

Combine $H^+ + OH^- \to H_2O$.

$$2\,H_2O + ClO^- \to Cl^- + H_2O + 2\,OH^-$$

Rewrite canceling H_2O on each side.

$$H_2O + ClO^- \to Cl^- + 2\,OH^-$$

Balance each half-reaction electrically with electrons.

$$6\,OH^- + I^- \to IO_3^- + 3\,H_2O + 6e^-$$
$$H_2O + ClO^- + 2e^- \to Cl^- + 2\,OH^-$$

Equalize loss and gain of electrons. Multiply reduction reaction by 3.

$$6\,OH^- + I^- \to IO_3^- + 3\,H_2O + 6e^-$$
$$3\,H_2O + 3\,ClO^- + 6e^- \to 3\,Cl^- + 6\,OH^-$$

Add the two half-reactions together, canceling the $6e^-$, $3\,H_2O$ and $6\,OH^-$ from each side of the equation.

$$3\,ClO^- + I^- \to IO_3^- + 3\,Cl^-$$

Check: each side of the equation has a charge of −4 and contains the same number of atoms of each element.

5. (1) No reaction. Copper is below lead on the series and therefore will not replace lead ions from solution.

   (2) Reaction. Magnesium is above zinc on the series and will therefore replace zinc ions from solution.

   $$Mg + ZnCl_2 \to Zn + MgCl_2$$

   (3) Reaction. Tin is above hydrogen on the series and will therefore replace hydrogen ions from solution. Reaction will occur.

   $$Sn + 2\,HCl \to SnCl_2 + H_2$$

   (4) Reaction. Aluminum is above silver on the series and will replace silver ions from solution.

   $$Al + 3\,AgNO_3 \to Al(NO_3)_3 + 3\,Ag$$

6. The Al^{3+} ions migrate toward the cathode, and the O^{2-} ions migrate toward the anode. To balance the two half-reactions, balance the electron gain and loss.

   $$Al^{3+} + 3e^- \to Al$$
   $$O^{2-} + C \to CO + 2e^-$$

   If you multiply the Al^{3+} half-reaction by 2 and the O^{2-} half-reaction by 3, the electrons will be balanced.

   $$2\,Al^{3+} + 6e^- \to 2\,Al$$
   $$3\,O^{2-} + 3\,C \to 3\,CO + 6e^-$$

   Added together: $2\,Al^{3+} + 3\,O^{2-} + 3\,C \to 2\,Al + 3\,CO$

7. (1) Molten $NiBr_2$ will exist as the following ions $NiBr_2 \xrightarrow{heat} Ni^{2+} + 2\ Br^-$. Oxidation, which is a loss of electrons, occurs at the anode. Br^- is the ion capable of losing electrons. The Ni^{2+} ion needs $2e^-$ (reduction) to become a Ni atom.

Therefore, anode reaction $2\ Br^- \rightarrow Br_2 + 2e^-$

The Ni^{2+} ion needs $2e^-$ (reduction) to become a Ni atom.

    Cathode reaction $Ni^{2+} + 2e^- \rightarrow Ni^0$

    Overall reaction $Ni^{2+} + 2\ Br^- \rightarrow Ni^0 + Br_2$

(2) LiCl will exist as Li^+ and Cl^- ions in the molten state. Oxidation at the anode will involve the Cl^- ion, but 2 Cl^- ions are needed to balance the equation since Cl_2 is produced.

    Anode $2\ Cl^- \rightarrow Cl_2 + 2e^-$

Reduction at the cathode will use the $2e^-$ available to reduce 2 Li^+ ions to lithium atoms.

    Cathode $2e^- + 2\ Li^+ \rightarrow 2\ Li^0$

    Overall $2\ Li^+ + 2\ Cl^- \rightarrow 2\ Li^0 + Cl_2$

8. (1) Fe^{2+} loses $1e^-$ to reach the Fe^{3+} oxidation state. Loss of e^- is oxidation, which occurs at the anode. Cr^{6+} goes from 6+ to 3+, which is a gain of electrons or reduction. This occurs at the cathode.

(2) In hydrogen peroxide (H_2O_2), oxygen is in a 1– oxidation state whereas in water oxygen is its usual 2–. This is reduction and occurs at the cathode. I^- loses electrons to reach the elemental state or zero oxidation state, I_2. This is oxidation or the anode reaction.

(3) Iron goes from the elemental state to 2+. This involves a loss of two elec-trons and takes place at the anode. Nickel is reduced from the 3+ state to 2+ state at the cathode.

These examples may be somewhat difficult. Look at them carefully and be sure you are clear on how to determine oxidation states of various atom. Refer to the text if necessary.

9. (1) voltaic   (2) electrolytic   (3) voltaic

10. Reaction (1) is oxidation, reaction (2) is reduction. Overall reaction:

    $Pb + PbO_2 + 4\ H^+ \rightarrow 2\ Pb^{2+} + 2\ H_2O$

11. 6.2 g/L

The problem is concerned with calculating the mass of chemical required for a "Normal" concentration solution. This means we have to determine the equivalent mass for $Na_2S_2O_3 \cdot 5\ H_2O$ for this particular reaction. First we must determine the chemical species in the $Na_2S_2O_3 \cdot 5\ H_2O$ that changes oxidation number and by how much. Looking at the reaction it appears that sulfur is involved.

Assigning oxidation numbers we find

    $Na_2S_2{}^{+2}O_3 \rightarrow Na_2S_4{}^{+2.5}O_6$

Each sulfur atom "loses" 1/2 e$^-$, but we see that each molecule of $Na_2S_2O_3 \cdot 5\ H_2O$ has 2 S atoms. Therefore, each molecule of $Na_2S_2O_3 \cdot 5\ H_2O$ loses 1e$^-$ which is the same as saying the equivalent for this reaction is the same as the molar mass.

Therefore,

$$0.025\ N\ \text{solution} = \left(0.025\ \frac{\cancel{\text{equiv}}}{L}\right)\left(248\ \frac{g}{\cancel{\text{equiv}}}\right) = 6.0\ g/L$$

12. (1) $MnO_4^- + 5\ VO^{2+} + H_2O \rightarrow 5\ VO_2^+ + 2\ H^+ + Mn^{2+}$
    (2) $P_4 + 3\ H_2O + 3\ OH^- \rightarrow PH_3 + 3\ H_2PO_2^-$
    (3) $MnO_2 + SO_3^{2-} + H_2O \rightarrow Mn(OH)_2 + SO_4^{2-}$
    (4) $5\ H_2O_2 + 2\ MnO_4^- + 6\ H^+ \rightarrow 2\ Mn^{2+} + 5\ O_2 + 8\ H_2O$
    (5) $2\ Mn^{2+} + 5\ HBiO_3 + 9\ H^+ \rightarrow 2\ MnO_4^- + 5\ Bi^{3+} + 7\ H_2O$
    (6) $Zn(s) + 2\ OH^-(aq) + 2\ H_2O(l) \rightarrow Zn(OH)_4^{2-}(aq) + H_2(g)$

13. (1) +6 (2) +2 (3) +4 (4) +4 (5) +3
    (6) +3 (7) +7 (8) +3 (9) +6 (10) +5

CHAPTER EIGHTEEN

Nuclear Chemistry

SELF-EVALUATION SECTION

1. Fill in the blank space or circle the correct response.

   In the symbol $^{80}_{27}X$, the atomic number of X is (1) _____ and the mass number of X is (2) _____ . The three principal rays or particles coming from the nucleus of a radioactive nuclide are (3) _____ , (4) _____ , and (5) _____ . When a radioactive nuclide loses a beta particle from the nucleus, its atomic number (6) increases/decreases by (7) _____ unit(s). At the same time, its atomic mass is essentially (8) the same/different . When a radioactive nuclide loses a/an (9) _____ particle, its mass number is (10) increased/ decreased by 4 amu, and its atomic number is (11) increased/ decreased by two units. The time it takes for one-half of a specific amount of a radioactive nuclide to disintegrate is called its (12) _____ . $^{223}_{88}Ra$ has a half-life of 11.2 days. If you have 8.0 grams of $^{223}_{88}Ra$ today, in (13) _____ days there will be 1.0 gram of the nuclide remaining.

2. The list below shows various terms related to the common types of radiation. Use it to construct the table as shown.

beta	4 amu	speed of light
gamma	0	rather slow speed
alpha	1 amu	rapid, 90% that of light
	-1	photons of light
$^{0}_{-1}e$	none	identical to He^{2+}
	1/1837 amu	identical to electron
-1		
$^{4}_{2}He$		
$+2$		

– 191 –

Symbol	Radiation Name	Mass (amu)	Charge	Velocity	Composition

3. Fill in the missing atomic number, atomic mass, or missing symbol for the following reactions. You may need to use a periodic table.

(1) $^{238}_{92}U \rightarrow {}^{234}Th + {}^{4}_{2}He$

(2) $^{}_{82}Pb \rightarrow {}^{214}_{83}Bi + {}^{0}_{-1}e$

(3) $^{210}_{84}Po \rightarrow {}^{206}_{82}Pb + \underline{}$

(4) $^{209}_{83}Bi + \underline{} \rightarrow {}^{210}_{84}Po + {}^{1}_{0}n$

(5) $^{35}_{17}Cl + {}^{1}_{0}n \rightarrow \underline{} + {}^{1}_{1}H$

(6) $\underline{} + {}^{1}_{0}n \rightarrow {}^{24}_{11}Na + {}^{4}_{2}He$

(7) $^{14}_{7}N + \underline{} \rightarrow {}^{14}_{6}C + {}^{1}_{1}H$

(8) $^{14}_{7}N + {}^{1}_{0}n \rightarrow {}^{12}_{6}C + \underline{}$

(9) $^{60}_{28}Ni + {}^{1}_{1}H \rightarrow {}^{57}_{27}Co + \underline{}$

(10) $^{7}_{3}Li + {}^{1}_{1}H \rightarrow \underline{} + {}^{1}_{0}n$

Write your answers here.

4. Identify the following reactions as nuclear fission or nuclear fusion reactions.

(1) $^{2}_{1}H + {}^{2}_{1}H \rightarrow {}^{3}_{2}He + {}^{1}_{0}n$

(2) $^{238}_{92}U + {}^{1}_{0}n \rightarrow {}^{144}_{56}Ba + {}^{90}_{36}Kr + 2\,{}^{1}_{0}n$

(3) $^{7}_{3}Li + {}^{1}_{1}H \rightarrow 2\,{}^{4}_{2}He$

(4) $^{2}_{1}H + {}^{2}_{1}H \rightarrow {}^{3}_{1}H + {}^{1}_{1}H$

(5) $^{1}_{0}n + {}^{238}_{92}U \rightarrow {}^{144}_{54}Xe + {}^{90}_{38}Sr + 2\,{}^{1}_{0}n$

(6) $^{3}_{1}H + {}^{2}_{1}H \rightarrow {}^{4}_{2}He + {}^{1}_{0}n$

5. Fill in the blank space or circle the correct response.

High levels of radiation, especially gamma or X rays, are termed (1) _____ radiation. If the dosage is high enough, (2) _____ can occur within several days. The effects of radiation appear to be localized in the (3) _____ of cells. Rapidly growing and dividing cells are (4) most/least susceptible to damage. Long-term or protracted exposure to (5) high/low levels of radiation can lead to health problems at some later time in a person's life. Presently there is concern about the effect that strontium-90, which is chemically similar to the element (6) _____ , has on blood cells manufactured in bone marrow. An accumulation of Sr-90 may lead to increased incidence of bone cancer and leukemia. Radiation damage to the nucleus of a cell can affect future generations of a particular species by giving rise to genetic (7) _____ . Such events occur when the radiation damages the genetic material – a molecule called (8) _____ – but not severely enough to prevent reproduction.

6. Match the names of scientists associated with nuclear chemistry with the appropriate descriptive phrase.

   (1) Becquerel
   (2) Marie Curie
   (3) Rutherford
   (4) E. O. Lawrence
   (5) Hahn and Strassman

   a. Discovered alpha and beta rays
   b. Developed cyclotron
   c. Coined word "radioactivity"
   d. Found that uranium salts emit rays
   e. Reported first instance of nuclear fission

7. Fill in the blank space or circle the correct response.

A magnetic field affects the three principal radioactive rays differently. The beta particle, being (1) positively/negatively charged, will be attracted (2) toward/away from the positive plate. The gamma ray, which has a mass of (3) _____ and an electrical charge of (4) _____ , (5) will/will not be attracted by the magnetic field. The alpha ray, which has a charge of (6) _____ , will be deflected (7) _____ the negative plate.

8. A piece of a wooden tool found at a Northwest Indian fishing village site has been determined to be approximately 10,000 years old. The ^{14}C in the wood has undergone approximately how many half lives of decay?

   Write your answer here.

9. The half life for $^{32}_{15}P$, a common biological radionuclide, is 14.3 days. Starting with 1500 micrograms of $^{32}_{15}P$, how much will you have left after 100 days?

   Do your calculations here.

10. Fill in the blank space or circle the correct response.

    The mass of an atomic nucleus is (1) <u>more</u>/less than the sum of the masses of the particles that make up the nucleus. The <u>difference</u> in mass is known as the (2) _____. The energy equivalent to this mass (using the equation $E = mc^2$) is called the (3) _____ of the nucleus. This amount of energy would be required to (4) put together/<u>pull apart</u> the particles of a particular nucleus. The higher the binding energy, the (5) <u>more</u>/ less stable the nucleus is. In both nuclear fission and fusion reactions, the products have less mass than the reactants. The resultant mass losses are ac-counted for in the very large quantities of energy that are released.

Challenge Problems

11. Calculate the binding energy for $^{32}_{16}S$ which occurs naturally at an abundance of 95%. The mass of one atom of $^{32}_{16}S$ is known to be 31.97207 amu (atomic mass units). The masses of the elemental particles are as follows:

    proton = 1.007277 amu   neutron = 1.008665 amu

    electron = 0.0005486 amu   1 amu = 1.49×10^{-10} J

    Do your calculations here.

12. The binding energy of one $^{7}_{3}$Li atom is 6.258×10^{-12} J. Using the masses of the elemental particles listed in problem 12, calculate the actual mass (in amu) of a $^{7}_{3}$Li atom.

$$1 \text{ amu} = 1.49 \times 10^{-10} \text{ J}$$

Do your calculations here.

RECAP SECTION

Chapter 18 is a fascinating chapter about a subject that influences us all. We are constantly bombarded by cosmic radiation, nuclear power plants pose problems of disposal of highly radioactive wastes, and we may sometime come into contact with radioactive isotopes during medical treatment. Radioactivity is not something to fear, but it is something to be discussed with knowledge and respect. Every citizen should know the essential details concerning radioactivity and its beneficial uses.

ANSWERS TO QUESTIONS AND SOLUTIONS TO PROBLEMS

1. (1) 27 (2) 80 (3) alpha (4) beta
   (5) gamma (6) increases (7) one (8) the same
   (9) alpha (10) decreased (11) decreased (12) half-life
   (13) 33.6

2. Check your table with Table 18.3 of the text.

3. (1) Atomic number of Th would be 90.

   (2) Mass number of Pb would be 214.

   (3) Missing species would be an alpha particle, $^{4}_{2}$He

   (4) Missing species would be an deuterium atom, $^{2}_{1}$H.

   (5) Missing species would be $^{35}_{16}$S. (6) Missing species would be $^{27}_{13}$Al.

   (7) Missing species would be $^{1}_{0}$n. (8) Missing species would be $^{3}_{1}$H.

   (9) Missing species would be $^{4}_{2}$H. (10) Missing species would be $^{7}_{4}$Be.

4. (1) Nuclear fusion — two light atoms combine to form a heavier nucleus.
    (2) Nuclear fission — heavy, unstable nucleus splits into two smaller fragments under bombardment with neutrons.
    (3) Nuclear fusion   (4) Nuclear fusion   (5) Nuclear fission

5. (1) acute   (2) death   (3) nucleus   (4) most
    (5) low   (6) calcium   (7) mutations   (8) DNA

6. (1) d   (2) c   (3) a   (4) b   (5) e

7. (1) negatively   (2) toward   (3) none   (4) zero
    (5) will not   (6) 2+   (7) toward

8. Approximately 2.

    The half-life for $^{14}_{6}C$ is 5668 years

9. $12\,\mu g$

    100 days is 7 half-lives. After this period of time 0.78% of the original material will be left.

    $0.0078 \times 1500\ \mu g = 12\ \mu g$

10. (1) less   (2) mass defect   (3) binding energy   (4) pull apart
    (5) more

11. Mass defect is equal to 0.29178 amu which is equal to 4.3475×10^{-11} J/atom of $^{35}_{16}S$.

12. 7.016 amu

    The calculated mass of the elemental particles in a $^{7}_{3}Li$ atom gives a value of 7.058137 amu. The binding energy is equal to 4.2×10^{-2} amu.

CHAPTER NINETEEN

Chemistry of Selected Elements

SELF-EVALUATION SECTION

1. True or False?
   (1) Most metals are found in nature in the form of cations. Processing the ore to produce free metal involves oxidation of the cations.
   (2) The higher the percent composition of gold in the alloy, the higher the carat.

2. A litmus solution turns blue when a small piece of sodium metal is placed in the solution. The metal disappears. Describe what has happened by means of a balanced equation.

3. Balance the following reactions by inspection or electron gain-loss method.
   (1) $Ca + U_2O_5 \xrightarrow{\Delta} U + CaO$

   (2) $Fe_2O_3 + CO \xrightarrow{\Delta} Fe_3O_4 + CO_2$

   (3) $Fe + HNO_3 \rightarrow Fe(NO_3)_2 + NH_4NO_3 + H_2O$

   (4) $Fe + H_2O \rightarrow Fe_3O_4 + H_2(g)$
       (Hot) (Steam)

   (5) $Al + H_2O \rightarrow Al_2O_3 + H_2(g)$

   (6) $Al + HCl \rightarrow AlCl_3 + H_2(g)$

   (7) $Al(OH)_3 + HCl \rightarrow AlCl_3 + H_2O$

4. In your own words, describe what is meant by cathodic protection of iron.

Next, from the list of metals given and referring to Table 18.3 in the text, identify by circling those which will give cathodic protection to iron.

        Cr   Al   Pb   Cu   Zn   Sn   Mg

5. Write the formulas for the following common compounds.

    (1) baking soda _____   (6) potash _____
    (2) magnesia _____   (7) limestone _____
    (3) quicklime _____   (8) gypsum _____
    (4) borax _____   (9) Epsom salts _____
    (5) lye _____   (10) soda ash _____

6. You are given a solid that is one of the following salts: $LiNO_3$, $NaNO_3$, KNO_3, $Sr(NO_3)_2$, or $Ba(NO_3)_2$. Outline a test you could easily perform to determine the identity of the salt and indicate the expected result.

7. What are the two biologically active alkaline earth metals?

8. List the essential materials needed for pig iron production and state the difference between pig iron and steel.

9. The preparation of pig iron from iron ore involves which of the following reactions?
   (1) $Fe + 6\ HNO_3 \rightarrow Fe(NO_3)_3 + 3\ NO_2(g) + 3\ H_2O$
   (2) $FeO + CO(g) \rightarrow Fe + CO_2(g)$
   (3) $2\ C + O_2(g) \rightarrow 2\ CO(g)$
   (4) $2\ Fe + 3\ Cl_2 \rightarrow 2\ FeCl_3$
   (5) $Fe_3O_4 + CO(g) \rightarrow 3\ FeO + CO_2(g)$
   (6) $CaO + SiO_2 \rightarrow CaSiO_3$
   (7) $CaCO_3 \rightarrow CaO + CO_2(g)$

10. Name the following binary compounds.
    (1) HBr _____
    (2) OF_2 _____
    (3) IBr _____
    (4) SO_2 _____
    (5) ICl_3 _____
    (6) CCl_4 _____

11. Why are alloys often used for industrial purposes in preference to a pure metal?

12. Determine the oxidation states of the elements in the following compounds.
    (1) ICl_3 _____
    (2) $HClO_2$ _____
    (3) S_2Cl_2 _____
    (4) Cl_2O_6 _____
    (5) H_2SO_4 _____
    (6) OF_2 _____
    (7) Cl_2O _____
    (8) BrF_5 _____

13. Balance the following equations and place correct coefficients in front of the chemical formulas.
    (1) $H_2S + H_2SO_4 \rightarrow S + H_2O$
    (2) $MnO_2 + HCl \rightarrow Cl_2 + MnCl_2 + H_2O$
    (3) $SiO_2 + HF \rightarrow SiF_4 + H_2O$

(4) $KClO_3 + HCl \rightarrow KCl + H_2O + Cl_2$

(5) $HBr + H_2SO_4 \rightarrow Br_2 + SO_2 + H_2O$

(6) $HNO_2 + HI \rightarrow I_2 + NO + H_2O$

(7) $HCl + HNO_3 \rightarrow Cl_2 + NO + H_2O$

(8) $H_2S + O_2 \rightarrow SO_2 + H_2O$

14. Identify the following reactions of sulfuric acid as either oxidizing agent, acid, or dehydrating agent.

(1) $NaOH + H_2SO_4 \rightarrow NaHSO_4 + H_2O$

(2) $C + 2\ H_2SO_4 \rightarrow CO_2(g) + 2\ SO_2(g) + 2\ H_2O$

(3) $Cu + 2\ H_2SO_4 \rightarrow CuSO_4 + 2\ H_2O + SO_2(g)$

(4) $C_{12}H_{22}O_{11} + 11\ H_2SO_4 \rightarrow 12\ C + 11\ H_2SO_4 \cdot H_2O$

(5) $CaCO_3 + H_2SO_4 \rightarrow CaSO_4 + CO_2(g) + H_2O$

15. Below are compounds that are used to prepare oxygen gas. Complete the reactions and balance.

(1) $HgO \xrightarrow{\Delta}$

(2) $H_2O \xrightarrow{electrolysis}$

(3) $H_2O_2 \rightarrow$

16. Arrange the halogens (except for astatine) in order of increasing strength as oxidizing agents.

17. Pick two examples of interhalogen compounds from question 12. Which element is considered electropositive?

18. Complete and balance the following equation, which represents a general method for preparing certain halogens.

$$NaCl + H_2SO_4 + MnO_2 \rightarrow MnSO_4 + Na_2SO_4 + (a) + (b)(g)$$

19. List essential materials needed for sulfuric acid production.

20. The preparation of sulfuric acid from sulfur involves which of the following reactions?
    (1) $2\ SO_2(g) + O_2(g) \xrightarrow{V_2O_5} 2\ SO_3(g)$
    (2) $8\ HI + H_2SO_4 \rightarrow 4\ I_2 + H_2S + 4\ H_2O$
    (3) $C + 2\ H_2SO_4 \xrightarrow{\Delta} CO_2(g) + 2\ SO_2(g) + 2\ H_2O$
    (4) $H_2SO_4 + SO_3(g) \rightarrow H_2S_2O_7$
    (5) $NaCl + H_2SO_4 \xrightarrow{\Delta} NaHSO_4 + HCl(g)$
    (6) $H_2S_2O_7 + H_2O \rightarrow 2\ H_2SO_4$

21. Write formulas for the following common names:
    (1) brimstone _____
    (2) fluorspar _____
    (3) Teflon _____
    (4) oleum _____
    (5) muriatic acid _____
    (6) bleach _____
    (7) aqua regia _____
    (8) saltpeter _____

22. Find the best match for each of the following medical uses of biological processes to the element largely involved.

Use/process	Element
(1) metal used in antiperspirants.	sodium
(2) helps prevent osteoporosis.	nitrogen
(3) plays a key role in skin integrity and healing wounds.	beryllium
(4) the carbonate salt is used to treat manic depression.	calcium
(5) essential component of ATP; makes up about 1% of the body's weight.	nitrogen
(6) used in X-ray tube windows.	zinc
(7) essential metal involved in the transport and storage of oxygen in the body.	zinc
(8) regulates pH in the body.	aluminum
(9) lightning flashes are one route by which this element is converted into compounds which are useful and necessary for higher forms of life.	phosphorus

23. Aluminum is produced by electrolysis of Al_2O_3. The balanced reaction is

$$Al_2O_3 + 3\ C \rightarrow 2\ Al + 3\ CO_{(g)}$$

How much aluminum is produced when 10 kg of Al_2O_3 is electrolyzed?

RECAP SECTION

Chapter 19 introduces the chemistry of important metallic and non metallic elements. Because metals are a non-renewable resource, we are learning very rapidly that re-cycling of steel and aluminum products is a necessity. Consequently, the science of metallurgy is expanding into many new areas. The text discusses the chemistry of several of the more important economic metals such as magnesium and aluminum, but all metals produced commercially have interesting chemical and physical properties. And even though the number of nonmetallic elements is small compared with metals, their importance is evidenced by considering that life depends on such elements as carbon, nitrogen, oxygen, sulfur, and phosphorus. The production of sulfuric acid (H_2SO_4) is said to indicate the overall condition of an industrialized economy, and ammonia is of almost equal importance. We will learn more about the biological importance of both non metallic and metallic elements in the succeeding chapters.

ANSWERS TO QUESTIONS AND SOLUTIONS TO PROBLEMS

1. (1) False. $M^+ \rightarrow M$ is a reduction. The cation must gain electrons to become the free metal.
   (2) True.

2. From the description in the question, a chemical reaction has taken place. The reactants are Na metal and H_2O. Hydroxide ions (OH^-) turn litmus blue. The suggestion is that they are produced from H_2O. The metal disappears and perhaps this means a soluble ion is produced. Let's put these ideas down on paper.

   $$Na^0 + H_2O \rightarrow Na^+ + OH^-$$

   We have accounted for most of the observed facts. However, notice that the sodium atom has lost an electron and has been oxidized. At the same time, another chemical element must be reduced in order to balance the electron gain and loss. The OH^- ion retains the same oxidation states as found in water. Therefore, we need to look elsewhere. A clue might be that the amount of hydrogen is not balanced. Can we use the hydrogen with an oxidation state of +1 from water to form a hydrogen molecule?

What must happen to an H^+ ion to be reduced? If two H^+ ions each gain one electron, they will form an H_2 molecule. This will lead us to a balanced equation.

$$2\ H^+ + 2e^- \rightarrow H_2(g)$$

$$2\ Na^0 \rightarrow 2\ Na^+ + 2e^-$$

Add in necessary H_2O and OH^- ions

$$2\ Na^0 + 2\ H_2O \rightarrow 2\ Na^+ + 2\ OH^- + H_2(g)$$

3. (1) $5\ Ca + U_2O_5 \xrightarrow{\Delta} 2\ U + 5\ CaO$
   (2) $3\ Fe_2O_3 + CO \xrightarrow{\Delta} 2\ Fe_3O_4 + CO_2$
   (3) $4\ Fe + 10\ HNO_3 \rightarrow 4\ Fe(NO_3)_2 + NH_4NO_3 + 3\ H_2O$
   (4) $3\ Fe + 4\ H_2O \rightarrow Fe_3O_4 + 4\ H_2(g)$
   (5) $2\ Al + 3\ H_2O \rightarrow Al_2O_3 + 3\ H_2(g)$
   (6) $2\ Al + 6\ HCl \rightarrow 2\ AlCl_3 + 3\ H_2(g)$
   (7) $Al(OH)_3 + 3\ HCl \rightarrow AlCl_3 + 3\ H_2O$

4. Cathodic protection of iron means using a metal higher on the activity series than iron to stop the iron from corroding or being oxidized. Thus, a metal higher on the list will be oxidized in preference to the iron, which in turn acts as the cathode rather than the anode. From the metals given, Cr, Al, Zn, and Mg will serve as anodes when attached to iron.

5. (1) $NaHCO_3$ (2) MgO (3) CaO
   (4) $Na_2B_4O_7 \cdot 10\ H_2O$ (5) $NaOH$ (6) K_2CO_3
   (7) $CaCO_3$ (8) $CaSO_4 \cdot 2\ H_2O$ (9) $MgSO_4 \cdot 7\ H_2O$
   (10) Na_2CO_3

6. The color of the flame in a flame test should easily distinguish these salts from each other. If $LiNO_3$, the flame should be red; if $NaNO_3$, a persistent yellow flame; if KNO_3, a short-lived violet flame; if $Sr(NO_3)_2$, a light green flame.

7. Ca and Mg

8. Iron ore, coke, and limestone are needed. Pig iron contains a number of impurities, including more carbon than steel does. Pig iron is converted to steel.

9. (2), (3), (5), (6), (7)

10. (1) hydrogen bromide   (4) sulfur dioxide
    (2) oxygen difluoride   (5) iodine trichloride
    (3) iodine monobromide   (6) carbon tetrachloride

11. The physical and chemical properties of the alloy are often more suitable for the industrial purpose. For example, depending on the metal alloyed with iron to make steel, steel is usually more resistant to corrosion, harder, tougher and stronger than iron. Other properties which are often improved by alloys include wearability, conductivity, tensile strength and high-temperature properties.

12. (1)   I = +3   (5)   H = +1
           Cl = −1   S = +6
                                       O = −2

     (2)   H = +1   (6)   O = +2
           Cl = +3   F = −1
           O = −2

     (3)   S = +1   (7)   Cl = +1
           Cl = −1   O = −2

     (4)   Cl = +6   (8)   Br = +5
           O = −2   F = −1

13. (1) $3\ H_2S + H_2SO_4 \rightarrow 4\ S + 4\ H_2O$
     (2) $MnO_2 + 4\ HCl \rightarrow Cl_2 + MnCl_2 + 2\ H_2O$
     (3) $SiO_2 + 4\ HF \rightarrow SiF_4 + 2\ H_2O$
     (4) $KClO_3 + 6\ HCl \rightarrow KCl + 3\ H_2O + 3\ Cl_2$
     (5) $2\ HBr + H_2SO_4 \rightarrow Br_2 + SO_2 + 2\ H_2O$
     (6) $2\ HNO_2 + 2\ HI \rightarrow I_2 + 2\ NO + 2\ H_2O$
     (7) $6\ HCl + 2\ HNO_3 \rightarrow 3\ Cl_2 + 2\ NO + 4\ H_2O$
     (8) $2\ H_2S + 3\ O_2 \rightarrow 2\ SO_2 + 2\ H_2O$

14. (1)   acid   (2)   oxidizing agent   (3)   oxidizing agent
     (4)   dehydrating agent   (5)   acid

15. (1) $2\ HgO \xrightarrow{\Delta} 2\ Hg + O_2(g)$
     (2) $2\ H_2O \xrightarrow{electrolysis} 2\ H_2(g) + O_2(g)$
     (3) $2\ H_2O_2 \rightarrow 2\ H_2O + O_2(g)$

16. F_2 is strongest, followed by Cl_2, Br_2, I_2.

17.   In compound (1), the I atom is positive; in compound (8), the Br atom is positive. These are the only two interhalogen compounds shown.

18.   Compound (a) is H_2O, compound (b) is Cl_2.
The balanced equation is

$2\ NaCl + 2\ H_2SO_4 + MnO_2 \rightarrow MnSO_4 + Na_2SO_4 + 2\ H_2O + Cl_2(g)$

19.   S, O_2, H_2SO_4, H_2O

20.   (1), (4), (6)

21. (1)   sulfur (S)   (2) CaF_2 (3) $(C_2F_4)_n$ (4) H_2SO_4
     (5)   HCl   (6)   NaOCl   (7) $HCl + HNO_3$ (8) KNO_3

22. (1)   aluminum   (2)   calcium   (3)   zinc   (4)   lithium
     (5)   phosphorus   (6)   beryllium   (7)   iron   (8)   sodium
     (9)   nitrogen

23. The problem is solved in the usual manner. We first find out the number of moles of Al_2O_3 that is represented by 10 kg and then relate this value to the number of moles of product or Al metal.

$$10 \text{ kg } Al_2O_3 = 10{,}000 \text{ g}$$

formula weight of $Al_2O_3 = 102 \ \dfrac{g}{mol}$

Therefore,

$$10{,}000 \text{ g} \times \dfrac{1 \text{ mol}}{102 \text{ g}} = 98.0 \text{ mol}$$

We know we are using 98.0 moles of Al_2O_3 as a reactant. How many moles of Al metal will we obtain from this amount? The equation states that 1 mole of Al_2O_3 yields 2 moles of Al^0.

$$Al_2O_3 + 3 \text{ C} \rightarrow 2 \text{ Al} + 3 \text{ CO}_{(g)}$$

Therefore,

$$\dfrac{2 \text{ mol Al}}{1 \text{ mol } Al_2O_3} \times 98.0 \text{ mol } Al_2O_3 = 196 \text{ mol Al}$$

$$\text{Number of grams} = 196 \text{ mol Al} \times 26.98 \dfrac{g}{mol}$$

$$= 5.29 \times 10^3 \text{ g Al (3 significant figures)}$$

CHAPTER TWENTY

Organic Chemistry: Saturated Hydrocarbons

SELF-EVALUATION SECTION

1. Fill in the blank space or circle the appropriate response.

   Carbon, with four outer shell electrons, is able to form (1) one/four single (2) ionic/covalent bonds by sharing its electrons with other elements. The bond angles are not planar but describe a (3) cubic/tetrahedronal shape. The remarkable ability of carbon to bond to (4) itself/only metals leads to the possibility of long chain compounds, ring compounds, and structures containing single, double, and even triple bonds. A single bond consists of (5) _____ electrons, a double bond of (6) _____ electrons, and a triple bond of (7) _____ electrons. For example, in the formula C_2H_4, each carbon atom, and the two hydrogen atoms are joined together by a (8) _____ covalent bond, and the two carbon atoms are joined together by a (9) _____ bond. The conventional diagrams of carbon to carbon bonds use two (10) vertical/horizontal dots to represent a single covalent bond, four dots for a double covalent bond and six dots for a triple covalent bond. The four outer shell electrons of carbon $(2s^2 2p^2)$ are hybridized during covalent bond formation to form three types of molecular orbitals. The most common type of molecular orbital is known as the sp^3 orbital. Four of these generate a tetrahedral structure about a central carbon nucleus.

   Since carbon can bond to itself, many organic compounds are "first cousins", or members of a homologous series. Each member of a homologous series differs from the others by a (11) fixed/varying weight. Thus, the alkanes or paraffins differ by only a CH_2 group or a fixed weight of 14.0 amu. Another interesting feature of organic compounds is the regular occurrence of compounds with the same molecular formula but different physical properties, such as boiling points and melting points. For example, the compounds normal butane and isobutane have the formula C_4H_{10} but have different properties. These compounds are called (12) _____ . The existence of isomers produced by the branching of carbon chains is one reason why there are so many organic compounds.

2. Circle the compounds or elements listed that **do not** contain carbon.

    (1) diamond
    (2) sugar
    (3) charcoal
    (4) silicon carbide
    (5) salt
    (6) graphite
    (7) coke
    (8) carbon black
    (9) plastics
    (10) acetic acid
    (11) ammonia
    (12) starch
    (13) carbon dioxide
    (14) sulfuric acid
    (15) fats
    (16) coal

3. Draw the Lewis electron-dot structures of carbon monoxide and carbon dioxide. Keep in mind the possibility of multiple bonds involving more than one pair of electrons.

4. Draw the Lewis electron-dot structures for butane and 3-chloropentane.

5. Match each of the following descriptions to one of the reactions below and then identify whether the reaction was substitution, elimination, or addition.

    Description

    (1) One atom is *removed* from a carbon which results in the carbon having a multiple bond to one of the remaining atoms.

    (2) One atom on carbon is *exchanged* for another.

    (3) One atom is *added* to carbon which previously was attached to less than four other atoms.

    Reaction

    (a) $CH_2=CH_2 + Cl_2 \rightarrow ClCH_2CH_2Cl$

    (b) $CH_3CH_2Cl + NaOCH_3 \rightarrow CH_2=CH_2 + NaCl + CH_3OH$

    (c) $CH_3CH_2Cl + LiBr \rightarrow CH_3CH_2Br + LiCl$

6. Match the structural formulas and the names for the isomers of hexane. Look for the longest continuous chain of carbon atoms and then the location of the alkyl groups along the chain.

   Formulas

   (1) CH$_3$—CH$_2$—CH$_2$—CH$_2$—CH$_2$—CH$_3$

   (2) CH$_3$—CH$_2$—CH$_2$—CH—CH$_3$
                                      |
                                    CH$_3$

   (3) CH$_3$—CH$_2$—CH—CH$_2$—CH$_3$
                        |
                      CH$_3$

   (4) CH$_3$—CH—CH—CH$_3$
                | |
             CH$_3$ CH$_3$

   (5)   CH$_3$
              |
       CH$_3$—C—CH$_2$—CH$_3$
              |
             CH$_3$

   Names

   3-methylpentane

   n-hexane

   2,3-dimethylbutane

   2-methylpentane

   2,2-dimethylbutane

7. Fill in the missing members of the alkane homologous series. Give the formulas and the names.

Alkanes			
methane	CH$_4$	_____	_____
ethane	C$_2$H$_6$	_____	_____
_____	_____	octane	C$_8$H$_{18}$
butane	C$_4$H$_{10}$	nonane	C$_9$H$_{20}$
pentane	C$_5$H$_{12}$	_____	_____

8. Name the following alkanes and alkyl halides. Look for the longest carbon chain and name them so that the substituent numbering is as small as possible.

   (1) CH$_3$—CH$_2$—CH—CH$_3$
                        |
                      CH$_3$

   (2)   CH$_3$ CH$_3$
                    | |
       CH$_3$—CH$_2$—C——CH—CH$_3$
                    |
                  CH$_3$

   (3)   Cl
                   |
       CH$_3$—CH$_2$—CH—CH$_2$—CH$_3$

(4) 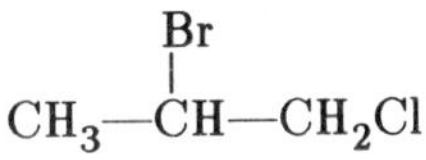 CH₃—CH(Br)—CH₂Cl _____

(5) 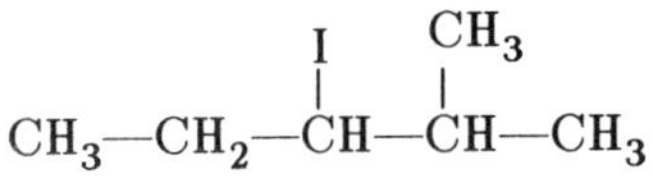 CH₃—CH₂—CH(I)—CH(CH₃)—CH₃ _____

(6) 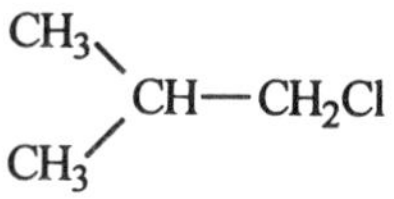 (CH₃)₂CH—CH₂Cl _____

(7) 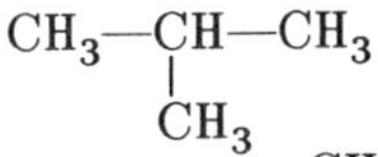 CH₃—CH(CH₃)—CH₃ _____

(8) 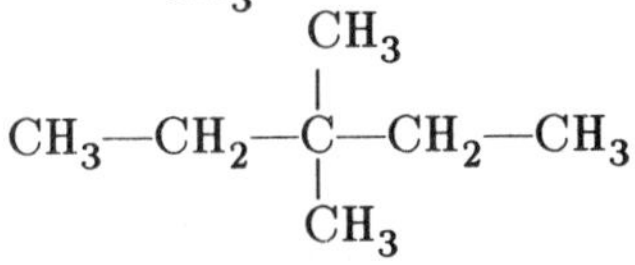 CH₃—CH₂—C(CH₃)₂—CH₂—CH₃ _____

9. The following molecules are isomers of chlorobutane. Identify the alkyl portions of the following molecules as n-butyl, sec-butyl, isobutyl, or tert-butyl.

   (1) 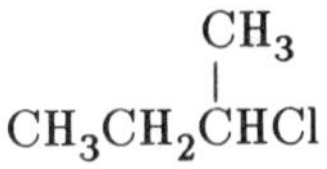 CH₃CH₂CHCl(CH₃)

   (2) 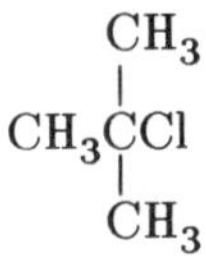 CH₃CCl(CH₃)CH₃

   (3) CH₃CH₂CH₂CH₂Cl

   (4) 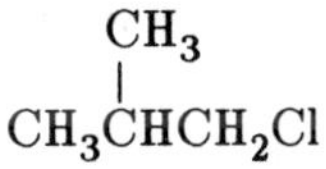 CH₃CH(CH₃)CH₂Cl

10. Draw the possible structures for the monochlorination of hexane. Eliminate any duplicate structures.

11. Oil companies produce gasoline more than most other petroleum products for use in the automobile industry.

    (1) How is gasoline a pollutant?

    (2) Why was lead a common additive to gasoline?

    (3) Why should a car requiring unleaded gasoline not use leaded gasoline even occasionally?

12. Either fill in the name or draw the structure for the following cycloalkanes.

Name	Structure
(1) _____	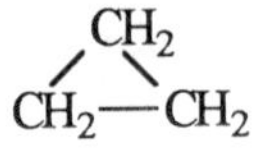 CH₂ / \ CH₂—CH₂
(2) cyclopentane	_____
(3) _____	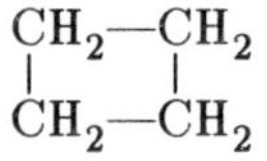 CH₂—CH₂ | | CH₂—CH₂
(4) cyclohexane	_____

13. What class of biochemical compounds possesses alkane-like properties?

RECAP SECTION

Chapter 20 has given you an introductory look at the chemistry of hydrocarbons, compounds composed entirely of carbon and hydrogen. In spite of the large number of possible molecules with single and double bonds and branching chains, organic chemists have developed a naming system that allows us to organize our knowledge. Today, organic chemists study reaction mechanisms, molecular structure, and the intricate synthesis schemes involving many types of organic molecules. However, the basic source of carbon compounds remains the hydrocarbons, coal and petroleum. Thus, in addition to being important for fuels, hydrocarbons are the beginning point for the other synthetic organic compounds in our lives.

ANSWERS TO QUESTIONS

1. (1) 4   (2) covalent   (3) tetrahedral   (4) itself
    (5) 2   (6) 4   (7) 6   (8) single
    (9) double   (10) vertical   (11) fixed   (12) isomers

2. Only (5), (11), and (14) do not contain carbon.

3. carbon monoxide   : C ::: O :

   carbon dioxide   : Ö :: C :: Ö :

   Carbon has four electrons and an oxygen atom has six. In order to write a correct Lewis electron-dot picture for carbon monoxide, CO, you need to have a triple bond between the two atoms. Two of the bonding electrons are from carbon and four are from oxygen.

Carbon dioxide with one carbon atom and two oxygen atoms involves all four of carbon's electrons in forming two double bonds. Each oxygen atom contributes two electrons to form the bonds. As you can see from the electron-dot diagrams for both molecules, each atom has eight electrons around it.

4. butane

$$H : \overset{H}{\underset{H}{C}} : \overset{H}{\underset{H}{C}} : \overset{H}{\underset{H}{C}} : \overset{H}{\underset{H}{C}} : H$$

3-chloropentane

$$H : \overset{H}{\underset{H}{C}} : \overset{H}{\underset{H}{C}} : \overset{\ddot{C}l}{\underset{}{}} : \overset{H}{\underset{H}{C}} : \overset{H}{\underset{H}{C}} : H$$

5. (1) (b) elimination (2) (c) substitution (3) (a) addition

6. (1) n-hexane (2) 2-methylpentane (3) 3-methylpentane
   (4) 2,3-dimethylbutane (5) 2,2-dimethylbutane

7. propane, C_3H_8; hexane, C_6H_{14}; heptane, C_7H_{16}; decane, $C_{10}H_{22}$

8. (1) 2-methylbutane (2) 2,3,3-trimethylpentane
   (3) 2-chlorobutane (4) 2-bromo-1-chloropropane
   (5) 3-iodo-2-methylpentane (6) 1-chloro-2-methylpropane
   (7) 2-methylpropane (7) 3,3-dimethylpentane

9. (1) sec-butyl (2) tert-butyl (3) n-butyl
   (4) isobutyl

10. There are six carbons in hexane so all the possible structures would be as follows (using a shorthand form):

$$\overset{6}{(Cl)}-\overset{|}{\underset{|}{C}}-\overset{\overset{5}{(Cl)}}{\underset{|}{\underset{|}{C}}}-\overset{\overset{4}{(Cl)}}{\underset{|}{\underset{|}{C}}}-\overset{\overset{3}{(Cl)}}{\underset{|}{\underset{|}{C}}}-\overset{\overset{2}{(Cl)}}{\underset{|}{\underset{|}{C}}}-\overset{1}{\underset{|}{\underset{|}{C}}}-(Cl)$$

There are six structures to look at for duplicates. Numbers 1 and 6 are identical since the molecule is the same when looked at from either end. Numbers 2 and 5 are identical, as are 3 and 4. Therefore, there are only three possible different monochlorination products of hexane.

$$CH_3-CH_2-CH_2-CH_2-CH_2-CH_2Cl$$
$$CH_3-CH_2-CH_2-CH_2-CHCl-CH_3$$
$$CH_3-CH_2-CH_2-CHCl-CH_2-CH_3$$

11. (1) Combustion of gasoline produces CO_2 thus increasing atmospheric CO_2 levels and contributing to the greenhouse effect.

(2) Tetraethyllead was added to gasoline to increase the octane rating (reduce knocking).

(3) Cars designed for unleaded gasoline are equipped with catalytic converters for the purpose of reducing pollutant emissions. Catalytic converters are deactivated by lead.

12. (1) cyclopropane

(2)
```
        CH₂
       /    \
    CH₂      CH₂
     |        |
    CH₂──────CH₂
```

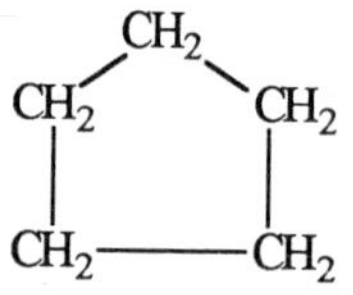

(3) cyclobutane

(4)
```
        CH₂
       /    \
    CH₂      CH₂
     |        |
    CH₂      CH₂
       \    /
        CH₂
```

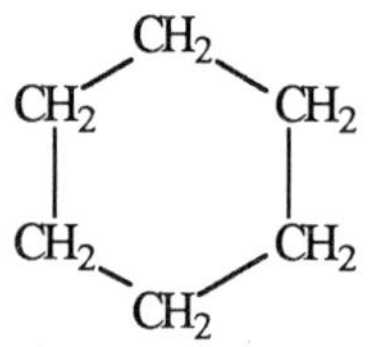

13. Fats and oils

CHAPTER TWENTY-ONE

Unsaturated Hydrocarbons

SELF-EVALUATION SECTION

1. Fill in blank space or circle the correct response.

   Alkenes and alkynes are classified as (1) saturated/unsaturated hydrocarbons since their molecules do not contain the maximum number of (2) _____ atoms possible. The alkenes contain at least (3) _____ double bond(s) and the alkynes contain at least (4) _____ triple bond(s) between carbon atoms. The formation of double and triple bonds between carbon atoms is explained by a hybridization process that differs from the sp^3 hybrid orbitals found in alkanes. For alkenes, the hybrid orbitals are termed (5) sp^2/sp. They are formed when one (6) _____ electron is promoted to a (7) _____ orbital to form four half-filled orbitals $2s^1\ 2p_x^{\ 1}\ 2p_y^{\ 1}\ 2p_z^{\ 1}$. Three of these orbitals hybridize to form three orbitals in a single plane with angles of (8) _____ between each other. The remaining $2p$ orbital is perpendicular to this plane with a lobe above and below the plane. During double bond formation between two carbon atoms, a hybridized sp^2 orbital from one carbon atom overlaps an identical sp^2 orbital from the second carbon atom. This produces a sigma bond. The perpendicular p orbitals on each carbon atom then overlap to form a pi bond. The pi bond with two electrons has two regions of electron cloud density—one above and one below the sigma bond. The remaining sp^2 orbitals on each carbon can form bonds to hydrogen atoms, halogen atoms, or other carbon atoms, for example. Alkynes with a triple bond have a bond formation process that is similar to that of alkenes. Two equivalent (9) sp^2/sp hybridized orbitals are formed with an angle between them of (10) _____. The remaining unhybridized $2p$ orbitals are oriented at right angles to each other and the hybridized sp orbitals. (11) _____ bonds are formed by the hybridized orbitals and two (12) _____ bonds are formed by overlap of the unhybridized p orbitals. In both alkenes and alkynes, the pi bonds are weaker than the sigma bonds, and therefore the molecules are very reactive.

2. Why is there no compound named methene or methyne?

3. Beginning with heptene, what are the next three members of the homologous series?

4. Alkanes have the general molecular formula C_nH_{2n+2}. What is the general molecular for alkenes? for alkynes?

5. Name the following compounds.
   (1) $CH_3-CH=CH_2$ _____
   (2) $CH_3-CH_2-CH_2-C\equiv CH$ _____
   (3) 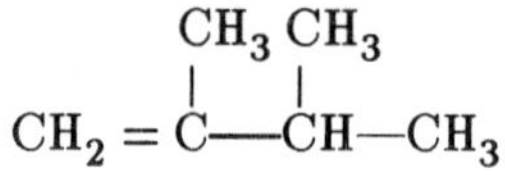 $CH_2=\underset{CH_3}{\overset{|}{C}}-\underset{CH_3}{\overset{|}{CH}}-CH_3$ _____
   (4) 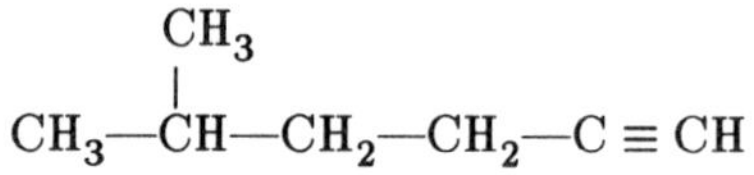 $CH_3-\underset{CH_3}{\overset{|}{CH}}-CH_2-CH_2-C\equiv CH$ _____
   (5) 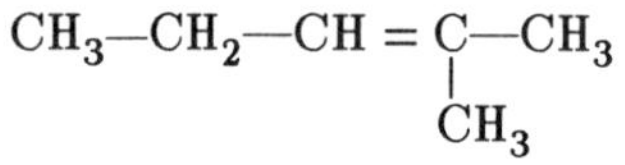 $CH_3-CH_2-CH=\underset{CH_3}{\overset{|}{C}}-CH_3$ _____
   (6) $CH_3-CH_2-CH=CH_2$ _____

6. Which of the following compounds can exist as geometric isomers?

   (1) 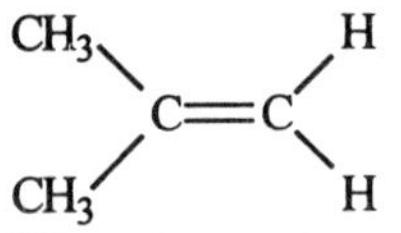 $\underset{CH_3}{\overset{CH_3}{\diagdown}}C=C\underset{H}{\overset{H}{\diagup}}$
   (2) $\underset{H}{\overset{CH_3}{\diagdown}}C=C\underset{H}{\overset{CH_3}{\diagup}}$
   (3) $CH_3-CH_2-\underset{CH_3}{\overset{|}{C}}=CH_2$
   (4) 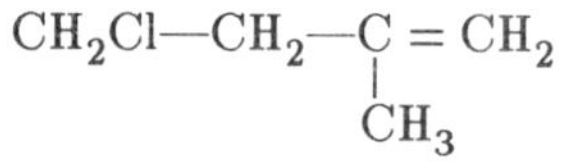 $CH_2Cl-CH_2-\underset{CH_3}{\overset{|}{C}}=CH_2$
   (5) $CCl_2=CHCl$
   (6) $CHCl=CHCl$

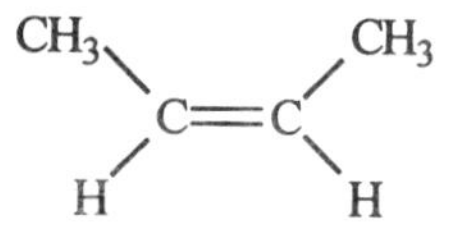

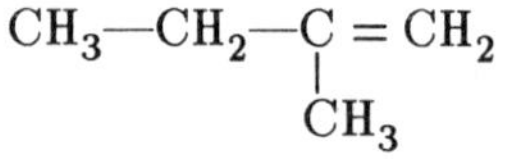

7. Draw the geometric structure of the following chemical compounds from their name.
   (1) cis-2-butene

   (2) trans-1,2-dichloroethene

   (3) cis-1-bromobutene

8. What is the hybridization of each carbon in the following molecule?

$(CH_3CH_2)_2C = CHCH_2C \equiv CH$

9. Predict the products formed in the following reactions. You may have to use Markovnikoff's rule.

(1) $CH \equiv CH + HCl \longrightarrow$

(2) $CH_2 = CH_2 + Br_2 \longrightarrow$

(3)
$$CH_3-\underset{\underset{CH_3}{|}}{C} = CH_2 + HI \longrightarrow$$

(4)
$$CH_3-CH_2-\underset{\underset{CH_3}{|}}{C} = CH-CH_3 + HBr \longrightarrow$$

(5) $CH_3-CH = CH-CH_3 + HI \longrightarrow$

(6) $CH_2 = CH_2 + H_2O \xrightarrow{H^+}$

10. Predict the color changes that will be observed during the following reactions.

(1) $CH_2 = CH_2 + Br_2 \longrightarrow$
   (reddish brown)

(2) $CH_3-CH = CH_2 + KMnO_4 + H_2O \longrightarrow$
   (purple)

11. Based on the chemical reactions in the last two chapters, how could you determine whether a 5 mole sample of an unknown compound is hexane, 2-hexene, or 2-hexyne?

12. What is the most common type of reaction alkenes and alkynes undergo? aromatics? All of these compounds have π bonds so why do they differ?

13. Name the aromatic compounds, using an appropriate system. You might choose to use either the numbering system or the o, m, p system if there are two substituents on the benzene ring.

(1) 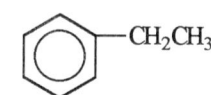 (2)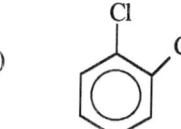

(3)  (4)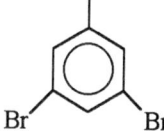

- 215 -

(5) 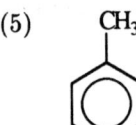　　　(6)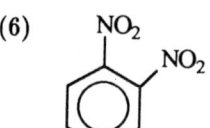

_____　　　_____

14. Draw the structures for the following compounds.

   (1) bromobenzene　_____

   (2) ethylbenzene　_____

   (3) phenol　_____

   (4) p-bromoaniline　_____

   (5) o-chloronitrobenzene　_____

   (6) 1,3,5-trinitrobenzene　_____

15. Match the correct fused aromatic ring structure with its name.

   (1) napthalene　　　　(a)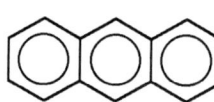

   (2) anthracene　　　　(b)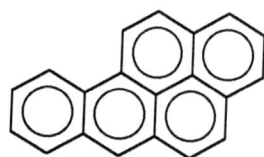

   (3) 3,4-benzpyrene　　(c)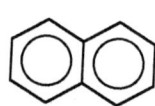

16. Complete the following reactions that involve aromatic hydrocarbons. Name each reaction type.

(1)

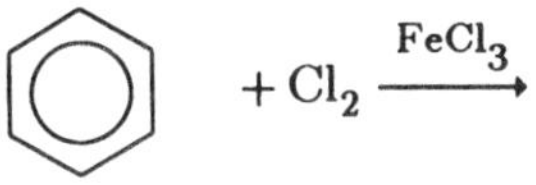

benzene + Cl_2 $\xrightarrow{FeCl_3}$

(2)

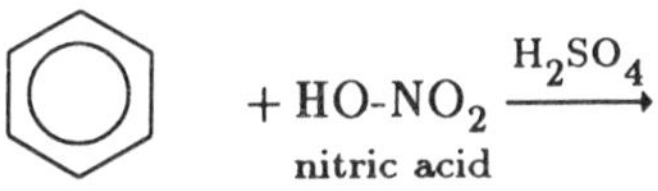

benzene + HO-NO$_2$ (nitric acid) $\xrightarrow{H_2SO_4}$

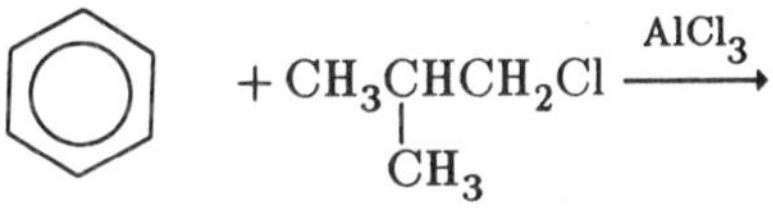

(3)

benzene + CH$_3$CHCH$_2$Cl $\xrightarrow{AlCl_3}$
             |
             CH$_3$

(4)

(benzene with CH$_2$CH$_3$ substituent) $\xrightarrow{KMnO_4/H_2O}$

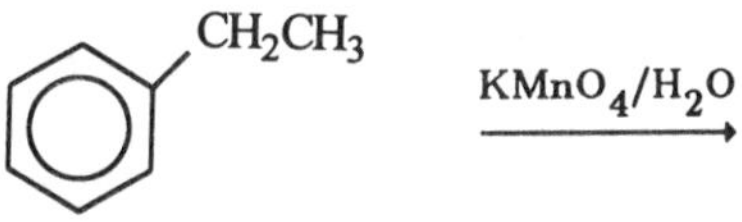

RECAP SECTION

Chapter 21 continues the discussion of hydrocarbon molecules begun in Chapter 20. Alkenes and alkynes are considerably more reactive than alkanes by virtue of the double and triple bonds. Aromatic hydrocarbons derived from the parent compound benzene exhibit a reactivity unlike that of the other hydrocarbons. Many important products in everyday use come from unsaturated hydrocarbon molecules. An understanding of these chemicals and their properties should be part of any chemistry student's learning.

ANSWERS TO QUESTIONS

1. (1) unsaturated (2) hydrogen (3) one (4) one
   (5) sp^2 (6) s (7) p (8) 120°
   (9) sp (10) 180° (11) sigma (12) pi

2. Alkenes and alkynes have carbon-carbon double bonds and triple bonds respectively and therefore a minimum of two carbons is required.

3. heptene C_7H_{14}, octene C_8H_{16}, nonene C_9H_{18}, decene $C_{10}H_{20}$

4. Alkenes: C_nH_{2n} Alkynes: C_nH_{2n-2}

5. (1) propene (2) 1-pentyne (3) 2,3-dimethyl-1-butene
   (4) 5-methyl-1-hexyne (5) 2-methyl-2-pentene (6) 1-butene

6. The compounds that can exist as geometric isomers are (2) and (6).

7. (1)
$$\begin{array}{c} CH_3 \\ \diagdown \\ H \end{array} C=C \begin{array}{c} CH_3 \\ \diagup \\ H \end{array}$$

   (2)
$$\begin{array}{c} Cl \\ \diagdown \\ H \end{array} C=C \begin{array}{c} H \\ \diagup \\ Cl \end{array}$$

   (3)
$$\begin{array}{c} CH_3-CH_2 \\ \diagdown \\ H \end{array} C=C \begin{array}{c} Br \\ \diagup \\ H \end{array}$$

8. $(CH_3CH_2)_2\underset{sp^2}{C}=CHCH_2\underset{sp}{C}\equiv CH$ The rest of the carbons are sp^3.

9. (1) $CH_2=CHCl$ (2) CH_2Br-CH_2Br

   (3) $CH_3-\underset{\underset{I}{|}}{\overset{\overset{CH_3}{|}}{C}}-CH_3$ (4) $CH_3-CH_2-\underset{\underset{Br}{|}}{\overset{\overset{CH_3}{|}}{C}}-CH_2-CH_3$

   (5) $CH_3-CH_2-\underset{\underset{I}{|}}{CH}-CH_3$ (6) CH_3-CH_2OH

10. (1) The reddish brown will disappear as the bromine adds to the double bond.
    (2) The purple color will become reddish brown. MnO_2 is formed.

11. To the sample, add a couple drops of bromine. If no reaction, the compound is hexane. If the color disappears, continue adding bromine until > 5 moles of Br_2 have been added. If the color persists, the sample was 2-hexene; if the color of Br_2 disappeared, the compound was 2-hexyne which needs 10 mol of Br_2 to fully react.

12. Alkenes and alkynes generally undergo addition reactions while aromatics usually do substitution reactions. The extra stability of the aromatic ring of π electrons is maintained by substitution while addition reactions would break the aromaticity.

13. (1) ethylbenzene
    (2) ortho-dichlorobenzene or 1,2-dichlorobenzene
    (3) para-xylene or 1,4-dimethylbenzene
    (4) 1,3,5-tribromobenzene
    (5) toluene or methylbenzene
    (6) ortho-dinitrobenzene or 1,2-dinitrobenzene

14. (1) 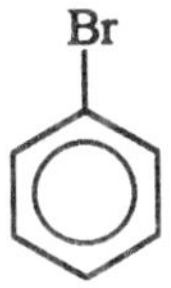 bromobenzene (Br on benzene ring)
    (2) 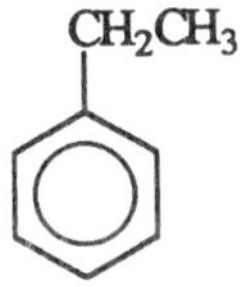 ethylbenzene (CH_2CH_3 on benzene ring)
    (3) phenol (OH on benzene ring)
    (4)  1,4-dibromo-... (NH_2 and Br para on benzene ring)
    (5) 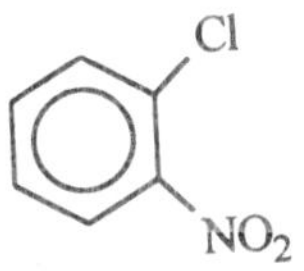 o-chloronitrobenzene (Cl and NO_2 on benzene ring)
    (6) 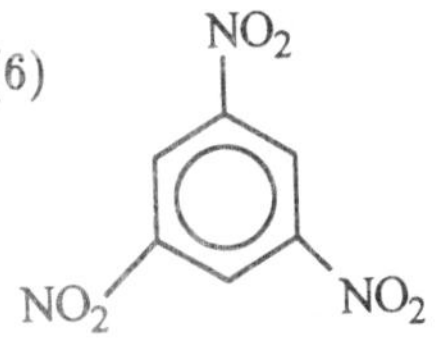 1,3,5-trinitrobenzene (three NO_2 groups on benzene ring)

15. (1) c   (2) a   (3) b

16. (1) 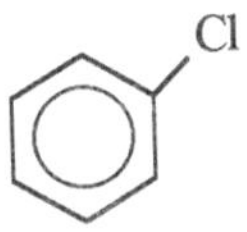 C$_6$H$_5$Cl + HCl — Halogenation
    (2) 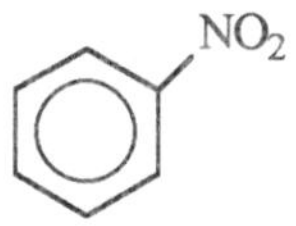 C$_6$H$_5$NO$_2$ + H$_2$O — Nitration
    (3) C$_6$H$_5$CH(CH$_3$)CH$_3$ [CH$_2$CHCH$_3$ with CH$_3$] 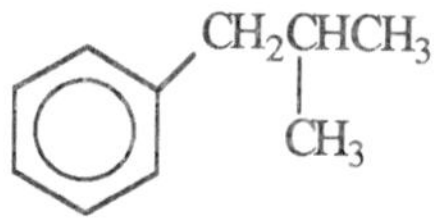 + HCl — Alkylation (Friedel-Craft)
    (4) 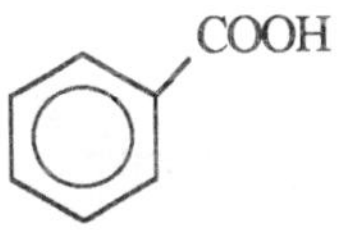 C$_6$H$_5$COOH + CO$_2$ — Oxidation

— 219 —

CHAPTER TWENTY-TWO

Alcohols, Ethers, Phenols, and Thiols

SELF-EVALUATION SECTION

1. Draw the structural formulas for the following alcohols.
   (1) 2-methyl-2-butanol
   (2) 3-chloro-1-butanol
   (3) 2,2-dimethyl-1-propanol
   (4) 3-methyl-2-butanol
   (5) 2,3-dimethyl-1-pentanol

2. Identify the following alcohols as primary, secondary, tertiary, or polyhydroxy.

   (1) 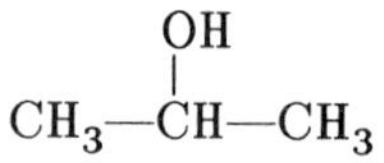 $\quad$ OH
   $\quad\quad$ |
   $\quad$ CH$_3$—CH—CH$_3$ $\quad$ _____

   (2) HO—CH$_2$—CH$_2$—OH $\quad$ _____

   (3) CH$_3$—OH $\quad$ _____

   (4) 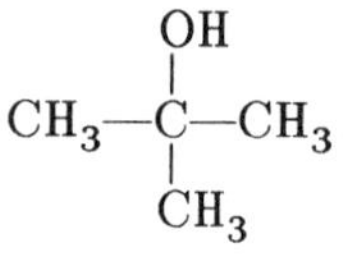 $\quad$ OH
   $\quad\quad$ |
   $\quad$ CH$_3$—C—CH$_3$
   $\quad\quad$ |
   $\quad\quad$ CH$_3$ $\quad$ _____

   (5) 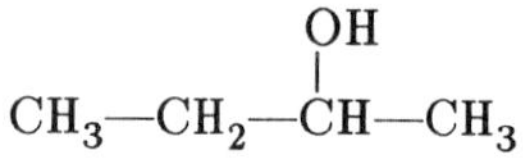 $\quad\quad\quad\quad$ OH
   $\quad\quad\quad\quad\quad$ |
   $\quad$ CH$_3$—CH$_2$—CH—CH$_3$ $\quad$ _____

   (6) 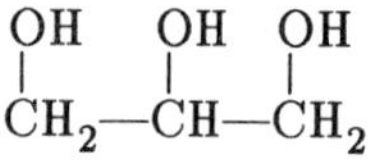 OH $\quad$ OH $\quad$ OH
   $\quad$ | $\quad\quad$ | $\quad\quad$ |
   $\quad$ CH$_2$—CH—CH$_2$ $\quad$ _____

   (7) 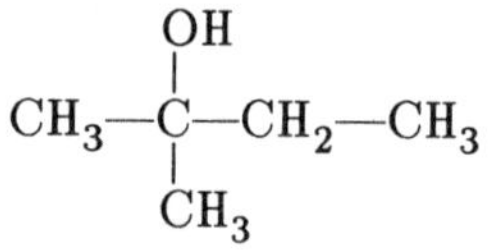 $\quad$ OH
   $\quad\quad$ |
   $\quad$ CH$_3$—C—CH$_2$—CH$_3$
   $\quad\quad$ |
   $\quad\quad$ CH$_3$ $\quad$ _____

   (8) CH$_3$CH$_2$CH$_2$CH$_2$OH $\quad$ _____

(9)
$$\text{CH}_3-\underset{\underset{\text{CH}_3}{|}}{\text{CH}}-\underset{\underset{\text{OH}}{|}}{\overset{\overset{\text{H}}{|}}{\text{C}}}-\text{CH}_3$$

_____

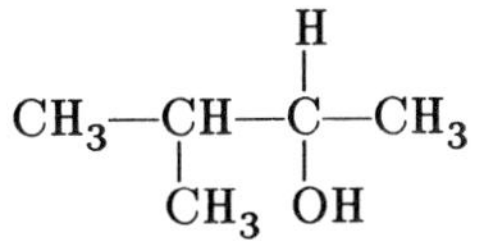

(10)
$$\text{CH}_3-\underset{\underset{\text{CH}_3}{|}}{\overset{\overset{\text{CH}_3}{|}}{\text{C}}}-\text{CH}_2\text{OH}$$

_____

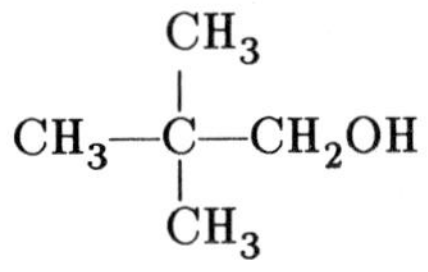

3. Name the first five alcohols of question 2.

4. Draw all possible isomers of the alcohol with the formula C_4H_9OH and name them.

5. The following section covers various reactions of alcohols. Decide whether a reaction will occur and what the product would be, and draw the chief product molecule and name it.

   (1) Oxidation with dichromate in sulfuric acid

      a. $CH_3CH_2OH \longrightarrow$

      b.
$$\text{CH}_3-\underset{\underset{}{}}{\overset{\overset{\text{OH}}{|}}{\text{CH}}}-\text{CH}_3 \longrightarrow$$

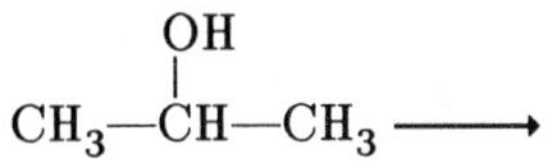

      c.
$$\text{CH}_3-\underset{\underset{\text{CH}_3}{|}}{\overset{\overset{\text{OH}}{|}}{\text{C}}}-\text{CH}_2-\text{CH}_3 \longrightarrow$$

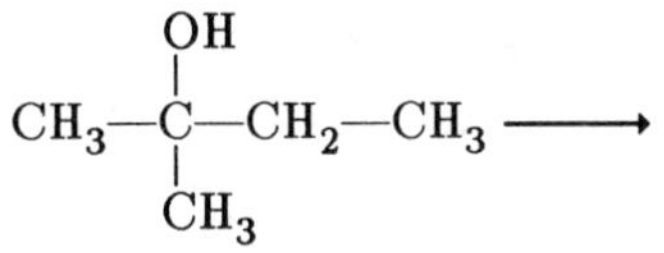

   (2) Dehydration with sulfuric acid

      a. $CH_3OH \longrightarrow$

      b.
$$\text{CH}_3-\overset{\overset{\text{OH}}{|}}{\text{CH}}-\text{CH}_3 \longrightarrow$$

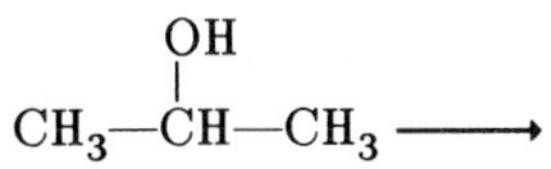

      c.
$$\text{CH}_3-\underset{\underset{\text{CH}_3}{|}}{\overset{\overset{\text{OH}}{|}}{\text{C}}}-\text{CH}_2-\text{CH}_3 \longrightarrow$$

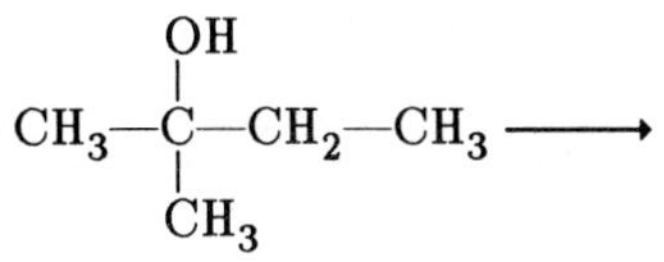

(3) Synthesis with alkyl halide

$$CH_3-CH_2-CH_2Cl + NaOH(aq) \longrightarrow$$

(4) Reaction of alcohols with sodium metal

$$2\ CH_3-CH_2-CH_2OH + 2\ Na \longrightarrow$$

6. The following are important common alcohols. Draw the structure underneath the name and list all possible descriptive phrases and terms that apply. There is some duplication.

Terms

(1) Methanol

(2) Ethanol

(3) Isopropylalcohol

(4) Ethylene glycol

(5) Glycerol

a. Made from ethylene
b. Made from propene
c. Ingredient for beverages
d. Conversion to formaldehyde
e. Wood alcohol
f. An industrial solvent
g. Permanent antifreeze
h. By-product of soap industry
i. Synthetic fiber manufacture
j. Used to make explosives
k. Poisonous
l. Grain alcohol
m. Used to make acetone
n. Made from carbon monoxide
o. From fermentation
p. Cosmetic manufacture

7. Fill in the blank space or circle the appropriate response.

   Ethanol is one of our oldest and best known (1) <u>ketones/alcohols</u>. The preparation by the process known as (2) _____ is still carried out today. The raw materials are usually (3) _____ and (4) _____ . A biological catalyst, called a(an) (5) <u>soap/enzyme</u>, is employed in natural fermentation to convert the raw materials into ethanol and carbon dioxide. Ethanol for industrial purposes is made from the two carbon alkene named (6) _____ and sulfuric acid. As a drug, ethanol has been shown to be a (7) <u>depressant/stimulant</u>, which is contrary to many people's beliefs. For nonfood uses, ethanol is made unfit for drinking by a process called (8) _____ . This process in effect poisons the ethanol.

8. What physical property could you use to determine whether a liquid is ethanethiol or ethanol? Explain why there is a difference.

9. Give the names for the following compounds.

   (1) $CH_3-CH_2-O-CH_2-CH_3$ _____

   (2) 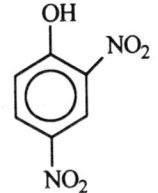 _____

   (3) $CH_3-O-CH_2-CH_3$ _____

   (4) 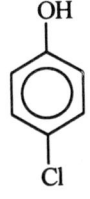 _____

   (5)

   _____

   (6) $CH_3-CH_2-O-\underset{\underset{CH_3}{|}}{CH}-CH_3$ _____

10. Phenols, which have an $-OH$ group attached directly to a (1) _____ ring, are weak (2) acids/bases . They will react with (3) sodium hydroxide/sodium bicarbonate solutions but not with (4) _____ . In general, phenols are (5) _____ to microorganisms and are (6) _____ to humans if ingested. Phenols used to be widely used as (7) _____ in hospitals, but now the primary use is in the manufacture of (8) _____ . Phenol can be derived from coal tar; however, a synthesis involving benzene and propene produces both phenol and (9) _____ . The intermediate, isopropyl-benzene, is also known as (10) _____ .

11. Rank the following molecules in increasing order of boiling point: 2-butanol, pentane, 2-butanethiol, 2-methoxypropane. Explain your choice.

12. Propose two different pathways to make $CH_3CH_2OCH_2CH_3$. Which pathway would be better to synthesize $CH_3OCH_2CH_3$? Why?

13. (1) Rank the following molecules in increasing order of expected solubility in water and explain your choice; diethylether, propanol, ethylene glycol, pentane.

    (2) Which of those molecules would be the best solvent for organic compounds?

RECAP SECTION

Chapter 22 has introduced us to some of the various functional groups that occur widely in organic compounds. Alcohols are among the most important compounds from an industrial standpoint and are an everyday part of our lives. Phenols and ethers play important roles also. The interrelationship of hydrocarbons, alcohols, aldehydes, and ketones is further discussed in Chapter 23. Many of the compounds mentioned in Chapters 22 and 23 are involved in biological reactions as well as in industrial processes.

ANSWERS TO QUESTIONS

1. (1) $CH_3CH_2-\underset{\underset{OH}{|}}{\overset{\overset{CH_3}{|}}{C}}-CH_3$

   (2) $CH_3\underset{\underset{Cl}{|}}{C}HCH_2CH_2OH$

   (3) $CH_3-\underset{\underset{CH_3}{|}}{\overset{\overset{CH_3}{|}}{C}}-CH_2OH$

   (4) $CH_3-\underset{\underset{}{|}}{\overset{\overset{CH_3}{|}}{C}H}-\underset{\underset{}{|}}{\overset{\overset{OH}{|}}{C}H}-CH_3$

   (5) $CH_3-CH_2-\underset{\underset{}{|}}{\overset{\overset{CH_3}{|}}{C}H}-\underset{\underset{}{|}}{\overset{\overset{CH_3}{|}}{C}H}-CH_2OH$

2. (1) secondary   (2) polyhydroxy   (3) primary
   (4) tertiary   (5) secondary   (6) polyhydroxy
   (7) tertiary   (8) primary   (9) secondary
   (10) primary

3. (1) 2-propanol or isopropanol   (2) 1,2-ethanediol or ethylene glycol
   (3) methanol   (4) 2-methyl-2-propanol or tert-butanol
   (5) 2-butanol

4. There are four possible isomers with a formula of C_4H_9OH:

(1) $CH_3CH_2CH_2CH_2OH$
1-butanol

(2) $CH_3-CH_2-\overset{\underset{|}{OH}}{C}HCH_3$
2-butanol

(3) $CH_3-\overset{\underset{|}{CH_3}}{\underset{|}{C}}-CH_3$ with OH on top
  2-methyl-2-propanol

(4) $CH_3-\underset{\underset{CH_3}{|}}{CH}-CH_2OH$
2-methyl-1-propanol

5. (1) a. Primary alcohol oxidized to an aldehyde and carboxylic acid.
    Products: ethanal CH_3CHO and acetic acid CH_3COOH.

   b. Secondary alcohol oxidized to ketone.
    Product will be acetone.
    $CH_3-\overset{O}{\overset{\|}{C}}-CH_3$

   c. Tertiary alcohol = no reaction

  (2) a. Primary alcohol dehydrates to ether.
    CH_3-O-CH_3 dimethyl ether

   b. Secondary alcohol dehydrates to alkene.
    $CH_3-CH=CH_2$ propene

   c. Tertiary alcohol dehydrates to alkene.
    $CH_3-\underset{\underset{CH_3}{|}}{C}=CH-CH_3$ 2-methyl-2-butene

  (3) Product will be an alcohol
   $CH_3CH_2CH_2OH$ propanol

  (4) Product will be a sodium alkoxide
   $CH_3CH_2CH_2ONa$ sodium propoxide

6. (1) Methanol   d, e, f, k, n
   CH_3OH

  (2) Ethanol   a, c, l, o, p
   CH_3CH_2OH

  (3) Isopropyl alcohol   b, f, k, m
   $CH_3\underset{\underset{OH}{|}}{CH}-CH_3$

  (4) Ethylene glycol   a, f, g, i, k
   $HO-CH_2-CH_2-OH$

(5) Glycerol   b, h, j, p

$$HOCH_2-\underset{\underset{OH}{|}}{CH}-CH_2OH$$

7. (1) alcohols   (2) fermentation   (3) starch
    (4) sugar   (5) enzyme   (6) ethylene
    (7) depressant   (8) denaturing

8. Ethanethiol can be differentiated from ethanol on the basis of smell or boiling point. Ethanethiol should smell like rotten eggs and because it does not hydrogen bond to itself like ethanol should have a lower boiling point than ethanol.

9. (1) diethyl ether or ethoxyethane
    (2) 2,4-dinitrophenol
    (3) methyl ethyl ether or methoxyethane
    (4) *p*-chlorophenol
    (5) *m*-methylphenol
    (6) ethyl isopropyl ether or 2-ethoxypropane

10. (1) benzene   (2) acids   (3) sodium hydroxide
    (4) sodium bicarbonate   (5) toxic   (6) poisonous
    (7) antiseptics   (8) resins and plastics
    (9) acetone   (10) cumene

11. Increasing order → Pentane, 2-methoxypropane, 2-butanethiol, 2-butanol.

Two factors to consider when estimating relative boiling points are molecular weight and strength of intermolecular forces. London dispersion foreces are weakest followed by dipole-dipole and then hydrogen bonding. Generally increasing molecular weight increases b.p. Therefore pentane with only LDF and similar mw to the other molecules has the lowest b.p. 2-butanol with H-bonding has the highest b.p. 2-butanethiol is slightly heavier and should have stronger dipole-dipole interaction than the ether and therefore should have a slightly higher b.p.

12. (1) $CH_3CH_2OH + CH_3CH_2OH \xrightarrow[140°C]{96\% H_2SO_4} CH_3CH_2OCH_2CH_3 + H_2O$

     (2) $CH_3CH_2ONa + CH_3CH_2Br \longrightarrow CH_3CH_2OCH_2CH_3 + NaBr$

The second pathway is best used to synthesize $CH_3OCH_2CH_3$ because a mixture of products would result if $CH_3OH + CH_3CH_2OH$ were used as reactants in the first pathway.

13. (1) Pentane, diethyl ether, propanol, ethylene glycol
        $CH_3CH_2CH_2CH_2CH_3$, $CH_3CH_2OCH_2CH_3$, CH_3CH_2OH, $HOCH_2CH_2OH$

The more polar a molecule and the more hydrogen-bonding capability of a molecule, the more soluble it will be in water.

     (2) Pentane would be the best solvent for a (non-polar) organic compound.

"Like dissolves like."

CHAPTER TWENTY-THREE

Aldehydes and Ketones

SELF-EVALUATION SECTION

The first three exercises contain material from Chapter 23 as well as earlier ones, and will help you to review alcohols, ethers, etc., as well as add the structural features of aldehydes and ketones to your knowledge.

1. Match the name of the functional group with the generalized formula.

   (1) alkane   _____   a. RX
   (2) aldehyde   _____   b. $\underset{\underset{R-C-R}{\|}}{O}$

   (3) alkyl halide _____   c. R—H
   (4) ether   _____   d. R—CH=CH$_2$
   (5) ketone   _____   e. R—OH
   (6) alcohol   _____   f. R—C≡CH
   (7) alkyne   _____   g. R—O—R
   (8) alkene   _____   h. R—CHO

2. After looking at the following list of structural formulas, match the names with various formulas. Some of the names will be IUPAC and some will be common names.

   (1) CH$_3$—O—CH$_3$ _____   a. ethanol
   (2) H$_2$C=O   _____   b. formaldehyde
   (3) CH$_3$CH$_2$OH   _____   c. benzaldehyde
   (4) $\underset{\underset{CH_3-C-CH_2CH_3}{\|}}{O}$ _____   d. t-butanol
   (5) $\underset{\underset{CH_3-CH-CH_3}{|}}{OH}$ _____
                                              e. dimethyl ether
   (6) HOCH$_2$CH$_2$OH   _____   f. 1,2-ethanediol or ethylene glycol
   (7) CH$_3$(CH$_2$)$_3$CH$_2$OH _____   g. 3-methyl-2-butanol

(8) 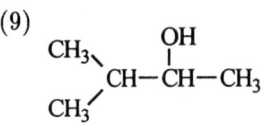   _____   h.   butanal

(9) 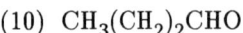   _____   i.   2-methylpropanal

(10) CH$_3$(CH$_2$)$_2$CHO   _____   j.   1-pentanol

(11)    _____   k.   methyl ethyl ketone

(12) CH$_3$—CH—CHO   _____   l.   isopropyl alcohol
          |
         CH$_3$

3. From the list of formulas given for question 2, identify each one as an alcohol, ether, aldehyde, or ketone.

   (1) _____   (2) _____
   (3) _____   (4) _____
   (5) _____   (6) _____
   (7) _____   (8) _____
   (9) _____   (10) _____
   (11) _____   (12) _____

4. How are all of the following molecules related to each other? Indicate the functional groups(s) in each molecule.

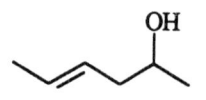 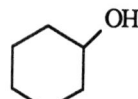 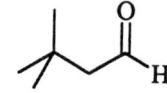 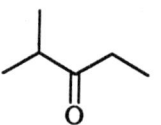 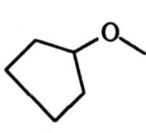

5. Can formaldehyde hydrogen-bond to itself (to other formaldehyde molecules)? to water? Which would you expect to have a higher boiling point: formaldehyde or methanol? Why?

6. Use the molecules below to answer the following questions.

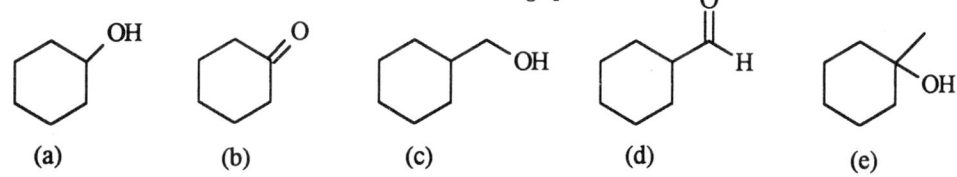

(a)   (b)   (c)   (d)   (e)

(1) Which molecules(s) is(are) easily oxidized with $H^+/Cr_2O_7^{2-}$?

(2) What bond is common to those molecules that is absent from the molecules which were not easily oxidized?

7. From the list of formulas given for question 2, identify the secondary alcohols and write out the ketone reaction product that would result from an oxidation reaction. Name the ketone product. Example:

$$CH_3-CH_2-\underset{\underset{\displaystyle OH}{|}}{CH}-CH_3 \xrightarrow{[O]} CH_3-CH_2-\underset{\underset{\displaystyle O}{||}}{C}-CH_3$$

2-butanol   2-butanone or methyl ethyl ketone

8. Write the reaction products for the following reduction reactions of aldehydes and ketones.

(1) Ph-CHO $\xrightarrow[H_2/Ni]{\Delta}$

(2) $CH_3-CH_2-\underset{\underset{\displaystyle O}{||}}{C}-CH_3 \xrightarrow[H_2/Ni]{\Delta}$

(3) $CH_3-CH_2-CHO \xrightarrow[H_2/Ni]{\Delta}$

9. Aldehydyes are easily oxidized to carboxylic acids by mild oxidizing agents. Some of the reactions are the bases of identifying tests. Give the product formed and the appearance for the following tests.

(1) Tollens' test

(2) Fehling's test

10. Identify which of the following molecules will undergo aldol condensation and write the products.

    (1) CH_3CHO

    (2) 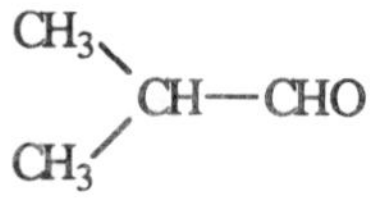 $(CH_3)_2CH-CHO$

    (3) $(CH_3)_3C-CO-C(CH_3)_3$

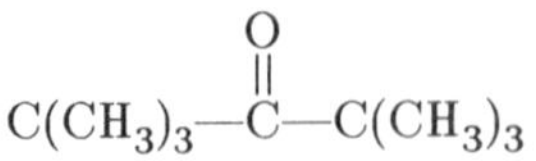

    (4) 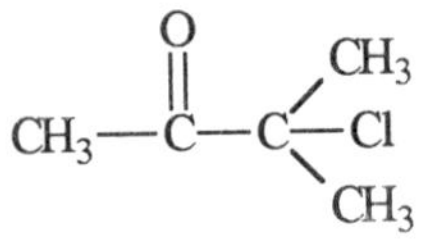 $CH_3-CO-C(CH_3)_2Cl$

11. Complete the following reactions for the formation of cyanohydrins.

    (1) 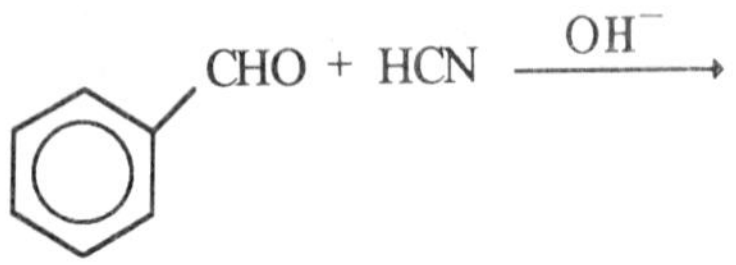 $C_6H_5-CHO + HCN \xrightarrow{OH^-}$

    (2) 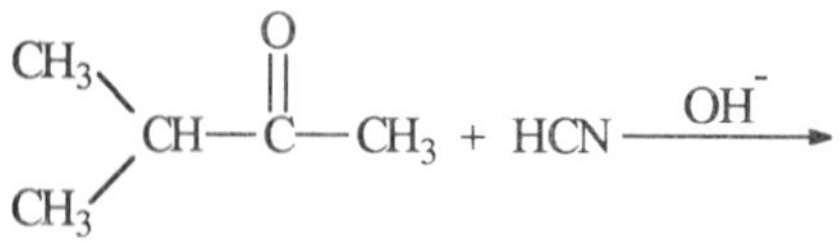 $(CH_3)_2CH-CO-CH_3 + HCN \xrightarrow{OH^-}$

12. Identify the following molecules as hemiacetals, hemiketals, acetals, or ketals.

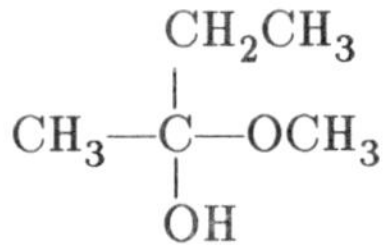

    (1) 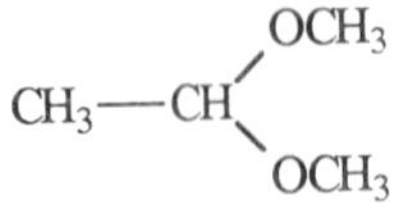 $CH_3-CH(OCH_3)_2$

    (2) $CH_3-C(CH_2CH_3)(OCH_3)(OH)$

    (3) $CH_3CH_2CH(OH)(OCH_2CH_3)$

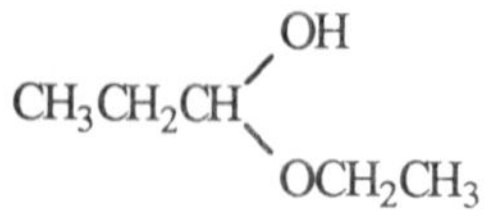

    (4) 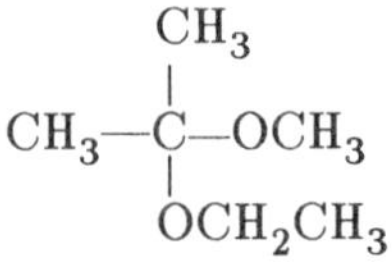 $CH_3-C(CH_3)(OCH_3)(OCH_2CH_3)$

13. How could you produce $CH_3CH_2CH_2CH(OH)COOH$ from $CH_3CH_2CH_2CHO$?

14. John was setting up a pair of experiments side by side. To one flask he added benzaldehyde and to the other he added acetaldehyde. Unknown to John, both flasks contained residual NaOH from a previous experiment. Before other reagents could be added, John was called away. Mary found the flasks sometime later and decided to quickly analyze their contents. What do you expect she discovered?

RECAP SECTION

Chapter 23 has introduced us to the chemistry of two more oxygen-containing compounds, aldehydes and ketones. We have seen the importance of the carbonyl group and learned about several different ways in which organic reactions take place. Numerous examples of important industrial and biological chemicals are mentioned in Chapter 23. The chemistry of sugars and carbohydrates is closely related to the concepts discussed in this chapter.

ANSWERS TO QUESTIONS

1. (1) c   (2) h   (3) a   (4) g
   (5) b   (6) e   (7) f   (8) d

2. (1) e   (2) b   (3) a   (4) k
   (5) l   (6) f   (7) j   (8) d
   (9) g   (10) h   (11) c   (12) i

3. (1) ether   (7) alcohol
   (2) aldehyde   (8) alcohol
   (3) alcohol   (9) alcohol
   (4) ketone   (10) aldehyde
   (5) alcohol   (11) aldehyde
   (6) alcohol (diol)   (12) aldehyde

4. The molecules are structural isomers of each other. The functional groups present are: alkene and alcohol, alcohol, aldehyde, ketone, ether.

5. $$H-\overset{\overset{O}{\|}}{C}-H$$

   Formaldehyde is unable to hydrogen-bond to itself but it should be able to hydrogen-bond to water (lone pairs on oxygen of formaldehyde with H atoms of water). Methanol can hydrogen-bond to itself and would therefore be expected to have a higher boiling point than formaldehyde. (The stronger the intermolecular forces, the higher the boiling point for molecules of similar molar mass.)

6. (1) The molecules which are easily oxidized are the 1° alcohol (c), the 2° alcohol (a), and the aldehyde (d). The ketone (b) and the 3° alcohol (e) cannot be easily oxidized.

   (2) All the molecules have a carbon bearing either a hydroxyl (OH) group or a double bond to oxygen. That carbon is also directly bonded to a hydrogen in the molecules which can be easily oxidized. Oxidations involving C—C breaking is not easily achieved. The C—H bond is always broken during these oxidations.

7. Secondary alcohols are (5) and (9)

   (5) $$CH_3-\overset{\overset{OH}{|}}{CH}-CH_3 \xrightarrow{[O]} CH_3-\overset{\overset{O}{\|}}{C}-CH_3$$

   2-propanone or acetone

(9)
$$\text{(CH}_3)_2\text{CH}-\underset{\text{OH}}{\text{CH}}-\text{CH}_3 \xrightarrow{[O]} \text{(CH}_3)_2\text{CH}-\underset{\text{O}}{\overset{\|}{\text{C}}}-\text{CH}_3$$
3-methyl-2-butanone

8. (1) C₆H₅–CH₂OH (benzyl alcohol)

   (2) $\text{CH}_3-\text{CH}_2-\underset{\text{OH}}{\text{CH}}-\text{CH}_3$

   (3) $\text{CH}_3\text{CH}_2\text{CH}_2\text{OH}$

9. (1) Tollens' test – silver mirror produced on test tube. Product is elemental metallic Ag.

   (2) Fehling's test – blue Cu^{2+} ions are reduced to Cu^+ and form brick red precipitate in test tube-Cu_2O-copper (I) oxide.

10. (1), (2), and (4) will undergo aldol condensations.

    (1) $\text{CH}_3\underset{\text{OH}}{\text{CH}}\text{CH}_2\text{CHO}$

    (2) $(\text{CH}_3)_2\text{CHCH}-\underset{\underset{\text{CH}_3}{|}}{\overset{\overset{\text{CH}_3}{|}}{\text{C}}}-\text{CHO}$
         with OH on the CHCH carbon

    (4) $\text{CH}_3-\underset{\underset{\text{CH}_3-\text{C}-\text{CH}_2-}{|}}{\overset{\overset{\text{CH}_3}{|}}{\text{C}}}-\text{Cl}$ ... $-\overset{\text{O}}{\overset{\|}{\text{C}}}-\underset{\underset{\text{CH}_3}{|}}{\overset{\overset{\text{CH}_3}{|}}{\text{CCl}}}$

11. (1) C₆H₅–CH(OH)(CN)

    (2) $(\text{CH}_3)_2\text{CH}-\underset{\underset{\text{CN}}{|}}{\overset{\overset{\text{OH}}{|}}{\text{C}}}-\text{CH}_3$

12. (1) acetal   (2) hemiketal   (3) hemiacetal
    (4) ketal

13. The carbon chain needs to be extended by one carbon. Addition of HCN to the aldehyde should produce a cyanohydrin which when hydrolyzed yields the carboxylic acid.

$$\text{CH}_3\text{CH}_2\text{CH}_2\overset{\text{O}}{\overset{\|}{\text{CH}}} \xrightarrow[\text{OH}^-]{\text{HCN}} \text{CH}_3\text{CH}_2\text{CH}_2\underset{\text{OH}}{\text{CHCN}} \xrightarrow[\text{H}_2\text{O}]{\text{H}^+} \text{CH}_3\text{CH}_2\text{CH}_2\underset{\text{OH}}{\text{CHCOOH}}$$

14. Residual NaOH can catalyze aldol condensations in aldehydes which contain an α-hydrogen. Benzaldehyde does not have α-hydrogens and therefore Mary should find benzaldehyde in the first flask. However, acetaldehyde does contain α-hydrogens and can undergo aldol condensation. Mary probably did not find acetaldehyde in the second flask—she most likely found 3-hydroxybutanol (which often continues reacting to form 2-butenal).

CHAPTER TWENTY-FOUR

Carboxylic Acids and Esters

SELF-EVALUATION SECTION

1. Name the organic acids using the IUPAC system.

(1) 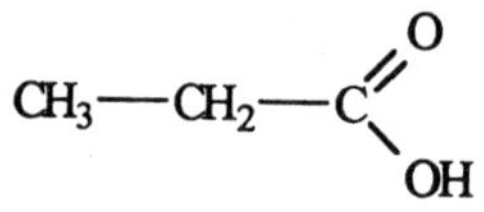

(4) 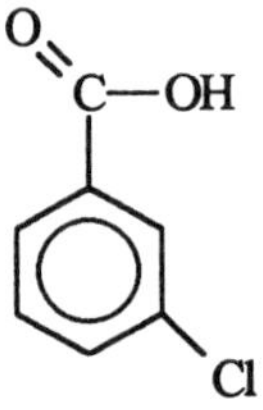

(2) 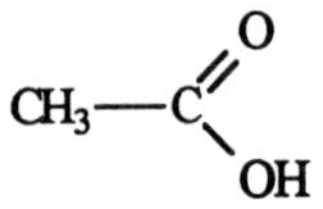

(5) 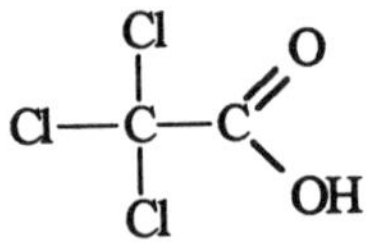

(3) 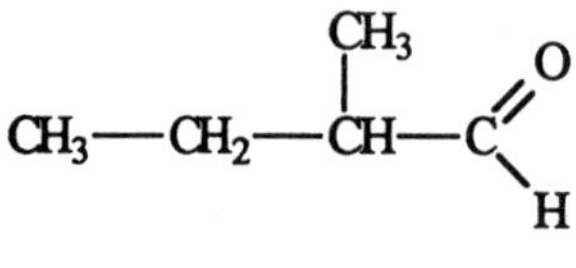

(6) 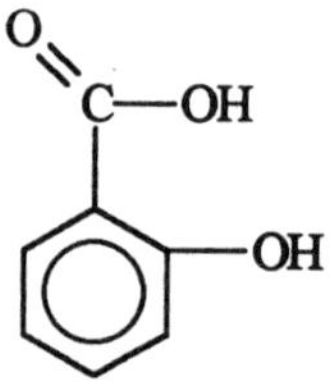

2. What is wrong with the following names:

(1) γ-hydroxypentanoic acid

(2) 3-aminocaprylic acid

3. Why does the water solubility of carboxylic acids decrease with increasing molecular weight?

4. Fill in the blank space or circle the correct response.

   Carboxylic acids show dramatic changes in their physical properties depending upon the number of carbon atoms in the molecule. The carboxyl group end is (1) polar/nonpolar and therefore (2) water soluble/insoluble. The carbon chain portion of the acid is (3) polar/nonpolar and therefore (4) water soluble/insoluble. For acids with one to four carbons, the (5) _____ end of the molecule takes precedence and the compound is (6) water soluble/insoluble. The solubility drops off rapidly after this number of carbons and long-chain fatty acids are insoluble.

5. Fill in the blank spaces with the chemical formula or the name of the following organic acids.

	Name	Formula
(1)	malonic acid	_____
(2)	_____	HOOC—COOH
(3)	_____	$CH_2=CHCOOH$
(4)	salicylic acid	_____
(5)	lactic acid	_____

6. Which three functional groups when directly bonded to an aromatic ring can be oxidized to produce benzoic acid?

7. Predict and name the organic acid product that will result from each reaction.

   (1) $CH_3OH \xrightarrow[H_2SO_4]{Cr_2O_7^{2-}}$ _____

   (2) $CH_3CH_2CH_2OH \xrightarrow[H_2SO_4]{Cr_2O_7^{2-}}$ _____

   (3) $\underset{\underset{CH_3}{|}}{CH_3-CH-CH_2OH} \xrightarrow[H_2SO_4]{Cr_2O_7^{2-}}$ _____

   (4) $\underset{\underset{CH_3}{|}}{CH_3-CH-CH_2-CHO} \xrightarrow[H_2SO_4]{Cr_2O_7^{2-}}$ _____

(5) 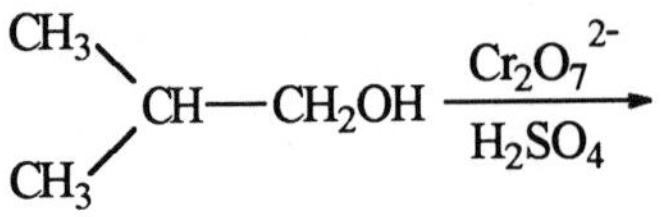 $\underset{CH_3}{\overset{CH_3}{>}}CH-CH_2OH \xrightarrow[H_2SO_4]{Cr_2O_7^{2-}}$ _____

(6) 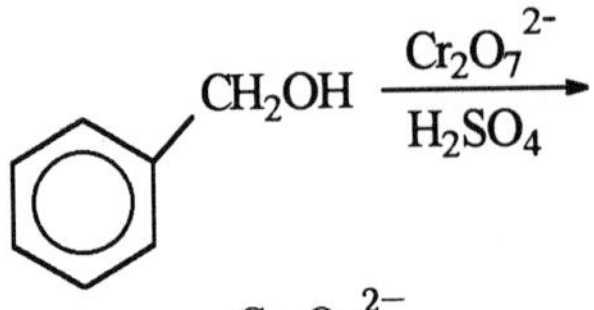 $C_6H_5-CH_2OH \xrightarrow[H_2SO_4]{Cr_2O_7^{2-}}$ _____

(7) $H_2C=O \xrightarrow[H_2SO_4]{Cr_2O_7^{2-}}$ _____

(8) 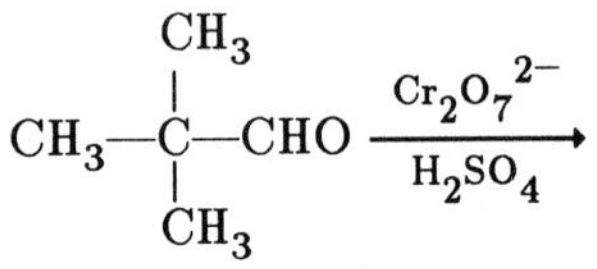 $CH_3-\underset{\underset{CH_3}{|}}{\overset{\overset{CH_3}{|}}{C}}-CHO \xrightarrow[H_2SO_4]{Cr_2O_7^{2-}}$ _____

8. Write the structural formulas and give the common names for:

    (1) Ethyl ethanoate

    (2) n-propyl methanoate

    (3) Methyl benzoate (no common name)

    (4) Isopropyl ethanoate

9. The following reactions involve carboxylic acids and their related compounds. Complete the reactions as called for.

    (1) Hydrolysis of a nitrile
    $$CH_3CN + 2\ H_2O \xrightarrow{H^+}$$

(2) Formation of a salt

$$CH_3COOH + NaOH \longrightarrow$$

(3) Formation of an ester

$$HCOOH + CH_3CH_2OH \xrightarrow{H^+}$$

(4) Formation of an acid chloride

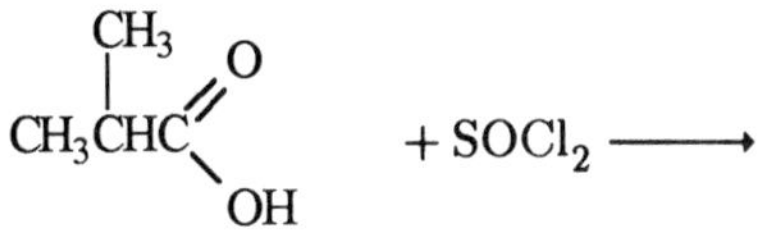

$$\underset{\underset{OH}{|}}{CH_3CHC} \overset{CH_3}{\underset{}{|}} \overset{O}{\underset{}{\parallel}} + SOCl_2 \longrightarrow$$

10. Name the following esters, using either the IUPAC system or a common name. Remember to name the acid portion of the ester last.

(1) 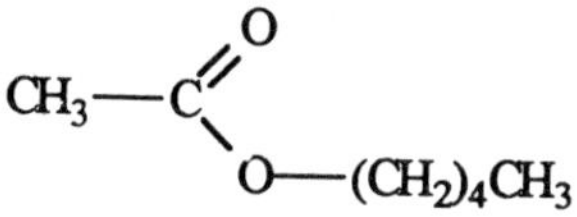 $CH_3-C(=O)-O-(CH_2)_4CH_3$ _____

(2) $CH_3-(CH_2)_2-C(=O)-O-CH_2CH_3$ _____

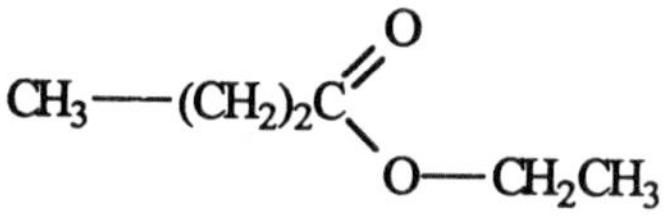

(3) $HC(=O)-O-CH_2-CH(CH_3)_2$ _____

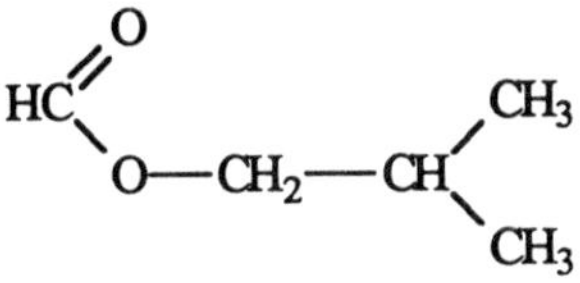

11. Predict the products of the following ester hydrolyses. Why does the pH of the hydrolysis make a difference in the products?

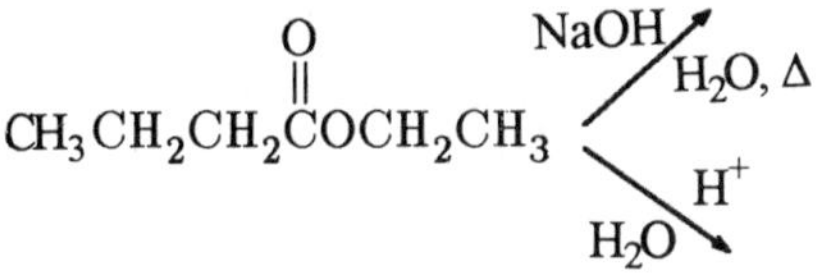

$$CH_3CH_2CH_2\overset{O}{\underset{\parallel}{C}}OCH_2CH_3 \quad \begin{array}{c} \xrightarrow{NaOH, H_2O, \Delta} \\ \xrightarrow{H^+, H_2O} \end{array}$$

12. Which member of each pair would you expect to have a higher boiling point? Why?

(1) propanol or acetic acid

(2) butanoic acid or ethyl acetate

(3) pentanic acid or sodium butanoate

13. Fill in the blank space or circle the appropriate response.

A chemical reaction as important today as when first utilized over 2000 years ago involves the saponification of a fat or oil with (1) acid/base. Fats and oils belong to a class of organic compounds called (2) esters/amides and usually contain (3) two/three molecules of organic acid esterified to a(n) (4) acid/polyhydroxy alcohol such as glycerol. The saponification reaction yields the salts of the organic acids, called (5) _____ , and the alcohol. Soaps are detergents and belong to a large class of compounds that have desirable cleansing and lubricating properties.

Soaps are derived from natural products obtained from animal and plant fats and basic compounds. Synthetic detergents (abbreviated (6) _____) began appearing on the market about 1930. Some of these syndets caused major pollution problems since they were not (7) _____ . For some time now, all of the manufactured detergents have been changed to a bio-degradable formula. One advantage a syndet has over a natural soap is that positive ions in hard water such as (8) _____ do not form insoluble precipitates with syndets.

A detergent molecule functions as cleansing agent by virtue of a nonpolar portion and a polar area in the molecule. A molecule such as sodium lauryl sulfate dissolves in water to produce two ions, (9) _____ and (10) _____ . The anion (11) _____ is the detergent species with its hydrocarbon tail, which is (12) nonpolar/polar . The (13) _____ portion dissolves in the greasy substances on soiled clothing, while the (14) _____ end interacts with the water molecules. The detergent species help form a stable emulsion between the oil or grease droplets and the water molecules so that mechanical agitation can lift the dirt from the clothing.

14. Match a term from the left with a description from the right.

    (1) Oleic acid   a. three molecules fatty acid + glycerol
    (2) Hydrogenolysis   b. salts of fatty acids + glycerol
    (3) Saponification   c. yields long chain alcohols + glycerol
    (4) Linoleic acid   d. one double bond
    (5) Hydrogenation   e. two double bonds
    (6) Linolenic acid   f. adds hydrogen to double bonds
    (7) Hydrolysis   g. three double bonds

15. Give the structure of the molecule that is formed when succinic acid is heated and a molecule of water is removed from the acid.

$$CH_2-C\begin{matrix}\nearrow O \\ \searrow OH\end{matrix}$$
$$CH_2-C\begin{matrix}\nearrow OH \\ \searrow O\end{matrix} \xrightarrow{\Delta}$$

succinic acid

16. Indicate the structural formulas and names of the products from the following acid hydrolysis reactions of esters.

    (1) $CH_3CH_2CH_2-\overset{\overset{O}{\|}}{C}-OCH_2CH_3 \xrightarrow{H^+}$

    (2) $CH_3-\overset{\overset{O}{\|}}{C}-O-\underset{}{\bigcirc} \xrightarrow{H^+}$

    (3) $H-\overset{\overset{O}{\|}}{C}-OCH_2CH_2CH_3 \xrightarrow{H^+}$

17. Show the structure of the product molecules after phosphoric acid is esterified with 3 molecules of ethanol.

    $HO-\overset{\overset{O}{\|}}{\underset{\underset{OH}{|}}{P}}-OH + 3\ CH_3CH_2OH \longrightarrow$

RECAP SECTION

Chapter 24 gives further examples of carbonyl containing compounds that play important roles in our everyday lives. Carboxylic acids and their derivatives have many functions in our bodies which will be explored in later chapters. Compounds containing these functional groups are major components of a wide variety of common chemicals. Paints, soaps, and flavorings are found in every household. The overcoming of the pollution problems brought about by the first syndets, a situation that was not suspected when the detergents came on the market, is an example of how chemical technology can provide solutions to today's problems.

ANSWERS TO QUESTIONS

1. (1) propanoic acid   (2) ethanoic acid or acetic acid
    (3) 2-methylbutanoic acid   (4) m-chlorobenzoic acid
    (5) trichloroethanoic acid   (6) o-hydroxybenzoic acid

2. Common names for acids should not be mixed with IUPAC nomenclature. The names should be:
    (1) γ-hydroxyvaleric acid *or* 4-hydroxypentanoic acid
    (2) β-aminocaprylic acid *or* 3-aminooctanoic acid

3. The carboxyl group is very polar, is able to participate in hydrogen-bonding, and is therefore highly soluble in water. As the molecular weight of the acid increases, the nonpolar R group of the acid dictates the solubility of the molecule therefore decreasing the solubility of the acid.

4. (1) polar (2) water soluble (3) nonpolar
   (4) insoluble (5) polar or carboxyl (6) water soluble

5. (1) COOH–CH$_2$–COOH (2) oxalic acid (3) propenoic acid
   (4) (2-hydroxybenzoic acid structure: benzene ring with COOH and OH)
   (5) CH$_3$CHCOOH with OH on middle carbon

6. A primary alcohol (C$_6$H$_5$CH$_2$OH), an aldehyde (C$_6$H$_5$CHO), and an alkyl group (e.g., C$_6$H$_5$CH$_2$CH$_2$CH$_3$) may all be oxidized to produce benzoic acid.

7. (1) methanoic acid or formic acid   HCOOH
   (2) propanoic acid   CH$_3$CH$_2$COOH
   (3) 2-methylpropanoic acid   CH$_3$–CH(CH$_3$)–COOH
   (4) 3-methylbutanoic acid   CH$_3$–CH(CH$_3$)–CH$_2$COOH
   (5) 2-methylpropanoic acid   CH$_3$–CH(CH$_3$)–COOH
   (6) benzoic acid   C$_6$H$_5$–COOH
   (7) formic acid or methanoic acid   HCOOH
   (8) 2.2-dimethylpropanoic acid   CH$_3$–C(CH$_3$)$_2$–COOH

8. (1) CH$_3$–C(=O)–OC$_2$H$_5$ ethyl acetate   (2) HC(=O)–O–CH$_2$CH$_2$CH$_3$ n-propyl formate

(3) $C_6H_5\text{—C(=O)—O—CH}_3$

(4) $CH_3\text{—C(=O)—O—CH(CH}_3)_2$ isopropyl acetate

9. (1) $CH_3COOH + NH_4^+$

   (2) $CH_3C(=O)ONa + H_2O$

   (3) $HC(=O)\text{—O—}CH_2\text{—}CH_3 + H_2O$

   (4) $CH_3\text{—CH(CH}_3)\text{C(=O)Cl} + SO_2 + HCl$

10. (1) pentyl ethanoate or pentyl acetate
    (2) ethyl butanoate
    (3) isobutyl formate or isobutyl methanoate

11. $CH_3CH_2CH_2\overset{O}{\overset{\|}{C}}OCH_2CH_3 \xrightarrow[\text{H}_2\text{O}, \Delta]{\text{NaOH}} CH_3CH_2CH_2\overset{O}{\overset{\|}{C}}ONa + CH_3CH_2OH$

$CH_3CH_2CH_2\overset{O}{\overset{\|}{C}}OCH_2CH_3 \xrightarrow[\text{H}_2\text{O}]{\text{H}^+} CH_3CH_2CH_2\overset{O}{\overset{\|}{C}}OH + CH_3CH_2OH$

When an alkaline hydrolysis of an ester is performed, the carboxylic acid that is produced is immediately deprotonated to produce the carboxylate salt (normal acid-base reaction). Alka-line hydrolysis is usually referred to as saponification.

12. Boiling points are highly dependent on molar mass and on the strengths of the inter-molecular forces. As both of those increase, so too does the boiling point.

    (1) acetic acid—carboxylic acids have more intermolecular hydrogen bonding (and usually form dimers) than alcohols
    (2) butanoic acid—esters (ethyl acetate) cannot hydrogen-bond to themselves
    (3) sodium benzoate—ionic bonds/interactions are much stronger than hydrogen-bonding

13. (1) base   (2) esters   (3) three
    (4) polyhydroxy alcohol   (5) soaps   (6) syndets
    (7) biodegradable   (8) Mg^{2+}, Ca^{2+} (9) Na^+
    (10) lauryl sulfate   (11) lauryl sulfate   (12) nonpolar
    (13) nonpolar   (14) polar

14. (1) d (2) c (3) b (4) e (5) f (6) g (7) a

15.

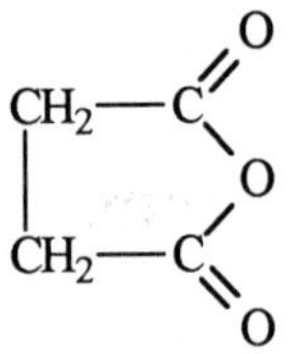

$$\begin{array}{c} CH_2-C\overset{\displaystyle O}{\underset{\displaystyle O}{\diagdown}} \\ | O \\ CH_2-C\underset{\displaystyle O}{\diagup} \end{array}$$

succinic anhydride

16. (1) $CH_3CH_2CH_2\overset{\displaystyle O}{\underset{\displaystyle \|}{C}}OH$ butyric acid or butanoic acid

CH_3CH_2OH ethanol

(2) $CH_3\overset{\displaystyle O}{\underset{\displaystyle \|}{C}}OH$ acetic acid

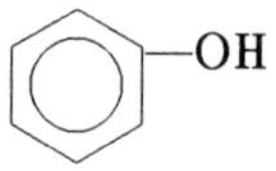

⟨◯⟩—OH   phenol

(3) $H\overset{\displaystyle O}{\underset{\displaystyle \|}{C}}OH$ formic acid

$CH_3CH_2CH_2OH$ propanol

17. $CH_3CH_2O-\underset{\displaystyle OCH_2CH_3}{\overset{\displaystyle \overset{O}{\|}}{P}}-OCH_2CH_3 + 3\,H_2O$

CHAPTER TWENTY-FIVE

Amides and Amines: Organic Nitrogen Compounds

SELF-EVALUATION SECTION

1. Complete the following reactions and name the amide that is formed in each case.

   (1) $CH_3CH_2\overset{\overset{O}{\|}}{C}-ONH_4 \xrightarrow{\Delta}$

   (2) $CH_3\overset{\overset{H_3C}{|}}{C}H\overset{\overset{O}{\|}}{C}-ONH_4 \xrightarrow{\Delta}$

   (3) 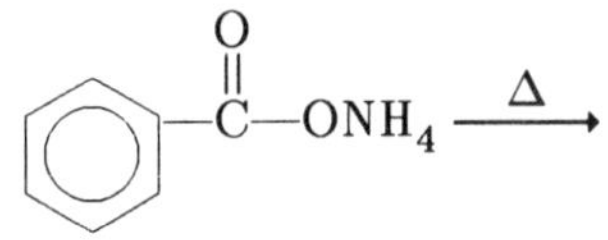 $\text{C}_6\text{H}_5-\overset{\overset{O}{\|}}{C}-ONH_4 \xrightarrow{\Delta}$

2. Give the products for acid and base hydrolysis of (1) propanamide and (2) benzamide.

3. Identify the following structural formulas as amines, amides, esters, or acids.

   (1) 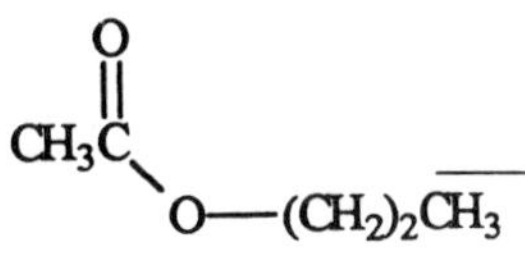 $CH_3\overset{\overset{O}{\|}}{C}-O-(CH_2)_2CH_3$ _____

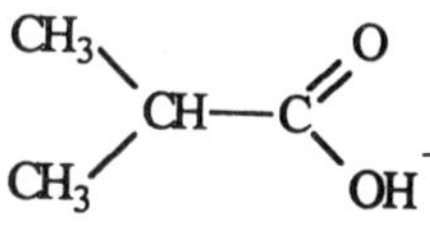

   (2) $CH_3-\underset{\underset{NH_2}{|}}{CH}-CH_3$ _____

   (3) $\overset{CH_3}{\underset{CH_3}{>}}CH-\overset{\overset{O}{\|}}{C}-OH$ _____

   (4) 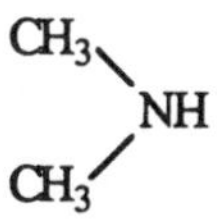 $\overset{CH_3}{\underset{CH_3}{>}}NH$ _____

(5) 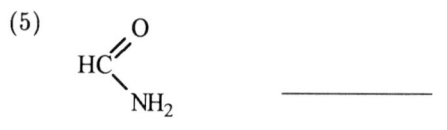 _____

(6) 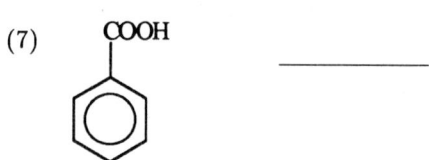 _____

(7) _____

(8) CH₃NH₂ _____

(9) 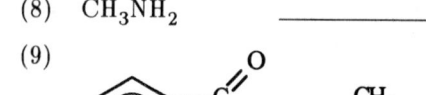 _____

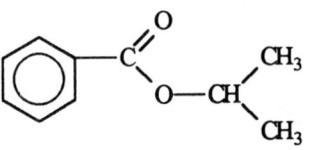

(10) _____

4. Write the formulas for the product ions when water reacts with the following amines and acids.

(1) $CH_3NH_2 + H_2O \longrightarrow$ _____

(2) $(CH_3)_2CH-COOH + H_2O \longrightarrow$ _____

(3) $C_6H_5NH_2 + H_2O \longrightarrow$ _____

(4) $C_6H_5OH + H_2O \longrightarrow$ _____

5. Identify the following amines as primary, secondary, or tertiary, and name them.

(1)

(2)

— 244 —

(3)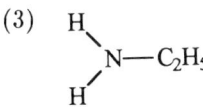

_____

(4)

_____

6. Which of the following would you expect to be the most soluble in water? the least soluble? Is there any way to induce the insoluble molecule to dissolve in water? Explain.

   (i) $(CH_3CH_2CH_2)_3 N$ (ii) CH_3NH_2 (iii) $CH_3\overset{\overset{O}{\|}}{C}NH_2$

7. Which of the molecules in the preceding question would have the highest boiling point? Why?

8. Amides are more difficult than esters to hydrolyse. Explain why this is important biologically.

9. Identify the following molecules as acidic, basic or neither.

   (a)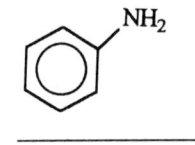

   (b)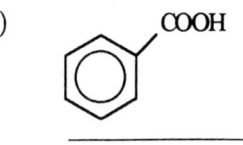

   (c)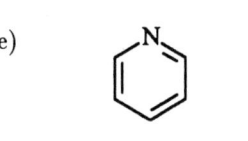

   _____   _____   _____

   (d) [structure]   (e) [structure]   (f) [structure]

   _____   _____   _____

- 245 -

10. Name the structures in (d) and (f) in the previous question. Are they identical molecules?

11. Many naturally occurring and synthetic compounds contain nitrogen. These drugs include amphetamines and barbituates. Give some specific examples of these compounds and their function.

12. Complete the following typical reactions for amines and indicate what type of reaction is occurring.

    (1) $CH_3NH_{2(g)} + HCl_{(g)} \longrightarrow$

    (2) $$CH_3\overset{\overset{O}{\|}}{C}-Cl + 2\,(CH_3CH_2)_2NH \longrightarrow$$

13. Complete the following reactions used to produce amines from amides and nitriles. What kind of reactions are these?

    (1) $$CH_3-\overset{\overset{O}{\|}}{C}-N\underset{CH_2CH_3}{\overset{CH_2CH_3}{\diagup\!\!\!\diagdown}} \xrightarrow{LiAlH_4}$$

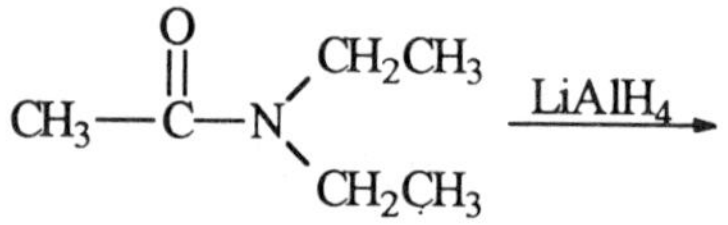

    N,N-diethylacetamide

    (2) $CH_3CH_2C \equiv N \xrightarrow{H_2/Pt}$

    propionitrile

14. What are two ways to prepare amines other than from amides or nitriles?

15. How would you separate a mixture of RCOOH, RNH_2, and ROH (where $R = C_7H_{15}$) based on solubility and abilities to act as an acid or a base?

RECAP SECTION

Chapter 25 discusses the two major classes of nitrogen-containing compounds—amines and amides. The properties, preparation, and reactions of these functional groups are introduced. These compounds are often important biological compounds, or, as synthetic compounds, are important commercial products. The sources and uses, many of which are drug-related, of some amines containing compounds are explored in this chapter. An understanding of nitrogen organic compounds is important in developing an understanding of molecular biochemistry and genetics.

ANSWERS TO QUESTIONS

1. (1) $CH_3CH_2\overset{O}{\overset{\|}{C}}-NH_2 + H_2O$ propanamide

   (2) $CH_3\overset{CH_3}{\overset{|}{C}}H\overset{O}{\overset{\|}{C}}-NH_2 + H_2O$ 2-methylpropanamide

   (3) $C_6H_5\overset{O}{\overset{\|}{C}}-NH_2 + H_2O$ benzamide

2. (1) Propanamide   acid $CH_3CH_2\overset{O}{\overset{\|}{C}}-OH + NH_4^+$

                           basic $CH_3CH_2\overset{O}{\overset{\|}{C}}-O^- + NH_3$

   (2) Benzamide   acid $C_6H_5\overset{O}{\overset{\|}{C}}-OH + NH_4^+$

                           basic $C_6H_5\overset{O}{\overset{\|}{C}}-O^- + NH_3$

3. (1) ester   (6) amine
   (2) amine   (7) acid
   (3) acid   (8) amine
   (4) amine   (9) ester
   (5) amide   (10) acid

4. (1) Product ions will be $CH_3NH_3^+ + OH^-$

   (2) $\begin{array}{c} CH_3 \\ \diagdown \\ CH-C \\ CH_3\diagup \diagup\diagdown \end{array} \begin{array}{c} O \\ \\ O^- \end{array} + H_3O^+$

   (3) [benzene ring]–$NH_3^+ + OH^-$

   (4) [benzene ring]–$O^- + H_3O^+$

5. (1) tertiary, trimethylamine   (2) primary, aniline
   (3) primary, ethylamine   (4) secondary, N-methylaniline

6. solubility: (iii) > (ii) > (i)   Solubility in water depends on how a molecule interacts with water. The more polar the molecule and the greater its ability to hydrogen bond, the more soluble the molecule. Amides are the most polar, therefore most soluble. If the tertiary amine is treated with an acid, the amine will be protonated and therefore changed/ionic. This now very polar molecule should dissolve in water.

7. The amide ($\overset{O}{\underset{\|}{CH_3CNHCH_3}}$) should have the highest boiling point. The amide bond is very polar as evidenced by the resonance structure and therefore there are electrostatic interactions between the amide molecules. The 1° amine has hydrogen bonding between molecules which is not as strong as the amide interactions. The 3° amine (i) should have the lowest boiling point as it cannot hydrogen bond to itself.

   (resonance structure of amide: $CH_3\overset{O^-}{\underset{}{C}}=\overset{}{\underset{H}{N^+}}-CH_3$)

8. Proteins are made up of amino acids which are linked by amide (peptide) bonds. If amides were easily hydrolyzed, our proteins would be too unstable.

9. (a) acidic   (b) basic   (c) acidic
   (d) neither (amides are not basic because the lone pair on nitrogen is unavailable due to resonance.)

   (e) basic   (f) neither

10. (d) acetanilide   (f) N-methyl benzamide   No, they are not identical.

11. Nicotine—tobacco derivative which stimulates the central nervous system
    Procaine—local anasthetic
    Nicotinamide—vitamin
    Methadone and cocaine—opiates

12. (1) $[CH_3NH_3]^+Cl^-$ acid-base reaction

(2) $CH_3\overset{\overset{O}{\|}}{C}-N\begin{smallmatrix}CH_2CH_3\\CH_2CH_3\end{smallmatrix}$ + $(CH_3CH_2)_2NH_2{}^+Cl^-$ condensation followed by acid-base

13. (1) $(CH_3CH_2)_3N$

(2) $CH_3CH_2CH_2NH_2$ These are reduction reactions.

14. Amines can be prepared from other amines by alkylation using an amine and an alkyl halide

example $CH_3CH_2Br + CH_3NH_2 \rightarrow CH_3CH_2NHCH_3$

Amines may also be prepared by reduction of nitrobenzene (useful for preparing aniline and aniline derivatives).

15. None of these molecules are likely to be soluble in water. If the mixture is treated with dilute acid (e.g., HCl), the amine will react to form an ammonium salt which should be soluble in water. Once that is separated (by extraction), add dilute base (e.g., $NaHCO_3$ or NaOH) which should react with the carboxylic acid. The carboxylate salt should now be soluble in water. To summarize, extract the organic solution successively with aqueous acid and then with aqueous base.

CHAPTER TWENTY-SIX

Polymers: Macromolecules

SELF-EVALUATION SECTION

1. Fill in the blank space or circle the appropriate response.

   The formation of very large molecules from smaller units is a process known as (1) _____ . The large molecule is called the (2) _____ and the small unit, the (3) _____ . For the polymer, polystyrene, the monomer unit is (4) _____ . Another term often used in reference to polymers is (5) micromolecule/macromolecule. The first synthetic polymers were made in the early part of the twentieth century. One particular one, discovered by Leo Baekeland in 1909 and called (6) _____ , was the first commercially successful polymer. It is still used for many applications such as electrical insulators and automobile parts. There are various ways to classify polymers. One way is based on the effect of temperature on a particular polymer. If the polymer softens on heating and can be reformed, it is said to be (7) _____ . On the other hand, a polymer that is hard, brittle, and will not soften when heated is a (8) _____ polymer. In these types of plastics, the polymeric chains are joined together or (9) _____ to form a network structure.

   Another classification is based on the manner in which the monomer units join together. When the monomers join together by a free radical reaction mechanism and produce a new longer chain-free radical, the polymer is called a(an) (10) _____ polymer. Other monomers have two reactive sites and join together with the splitting out of a small molecule such as H_2O. The chain can grow at either end. This type of polymer is called a(an) (11) _____ polymer. Polyethylene, PVC, Orlon, Teflon, and Lucite are all examples of (12) _____ polymers, while Dacron, Nylon, Bakelite, and polyurethanes are examples of (13) _____ polymers. The addition polymer, natural rubber, is usually vulcanized during its manufacture into various products. The process invented by (14) _____ in 1839 introduces (15) _____ atoms that (16) _____ the long polymeric chains.

2. Label the following reactions as chain initiating, chain propagating, or chain terminating and then order the steps.

(1) 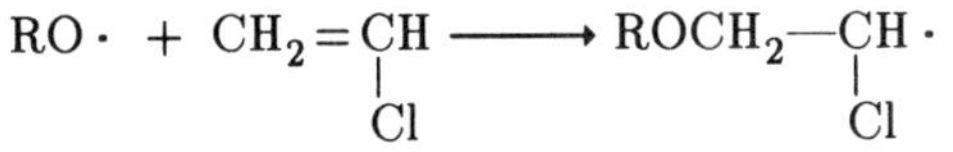 RO· + CH$_2$=CH(Cl) ⟶ ROCH$_2$—CH·(Cl)

_____

(2) ROCH$_2$—CH(Cl)—CH$_2$—CH·(Cl) + RO· ⟶ ROCH$_2$—CH(Cl)—CH$_2$—CHOR(Cl)

_____

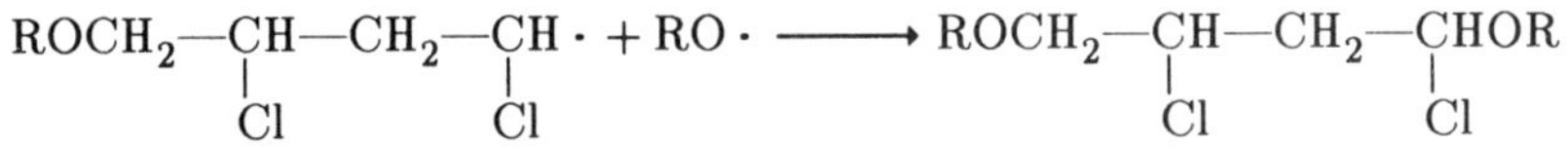

(3) RO:OR ⟶ 2 RO·

_____

(4)  ROCH$_2$—CH·(Cl) + CH$_2$=CH(Cl) ⟶ RO—CH$_2$—CH(Cl)—CH$_2$—CH·(Cl)

_____

(5) 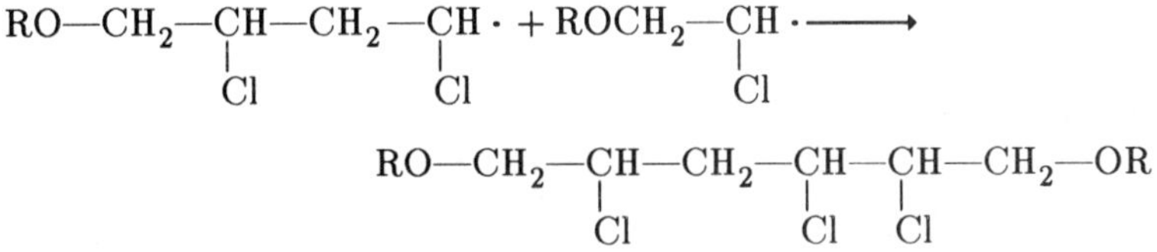 RO—CH$_2$—CH(Cl)—CH$_2$—CH·(Cl) + ROCH$_2$—CH·(Cl) ⟶
     RO—CH$_2$—CH(Cl)—CH$_2$—CH(Cl)—CH(Cl)—CH$_2$—OR

_____

3. Match the monomer molecules listed in the left-hand column with the generalized formulas for the corresponding polymers in the right-hand column. There may be more than one monomer associated with a polymer, as in a copolymer. For any polymers that are not matched, draw a possible monomer.

(1) 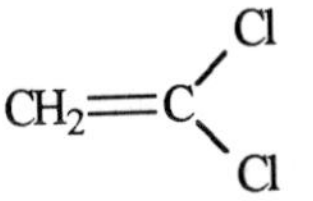 CH$_2$=C(Cl)(Cl)   _____   a. $(CF_2-CF_2)_n$

b. 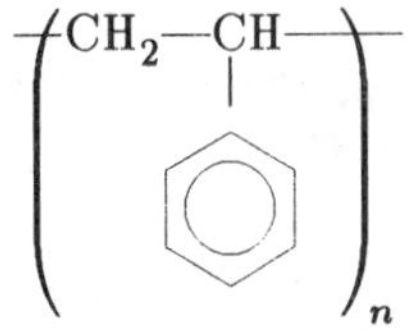 $(CH_2-CH(C_6H_5))_n$

— 251 —

(2) $CH_2=CH$
     $\quad\ \ \ |$
     $\quad\ \ CN$

(3) 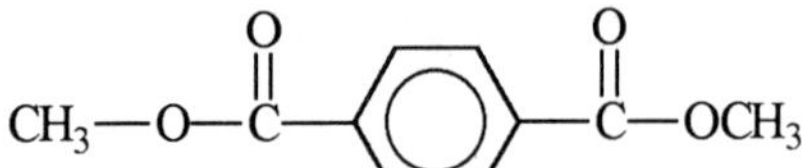 $CH_3-O-\overset{O}{\underset{\|}{C}}-\bigcirc-\overset{O}{\underset{\|}{C}}-OCH_3$

(4) $\quad\ \ CH_3\ O$
     $\quad\ \ \ \ |\ \ \ \ \|$
     $CH_2=C-C-OCH_3$

(5) 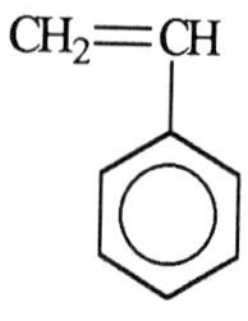 $CH_2=CH-\bigcirc$

(6) $CF_2=CF_2$

(7) $CH_2=C\overset{H}{\underset{CH_3}{\diagdown}}$

(8) $HOCH_2CH_2OH$

c. $-(CH_2-\underset{Cl}{\overset{Cl}{\underset{|}{\overset{|}{C}}Cl_2}})_n$

d. $-(CH_2-\underset{COOCH_3}{\overset{CH_3}{\underset{|}{\overset{|}{C}}}})_n$

e. $-(CH_2-\underset{CH_3}{\overset{H}{\underset{|}{\overset{|}{C}}}})_n$

f. 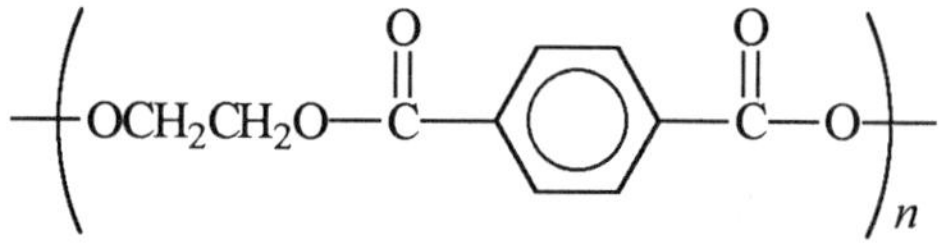 $-(OCH_2CH_2-O-\overset{O}{\underset{\|}{C}}-\bigcirc-\overset{O}{\underset{\|}{C}}-O)_n$

g. $-(CH_2-\underset{CN}{\overset{}{\underset{|}{CH}}})_n$

h. $-(CH_2-\underset{CH_3}{\overset{CH_3}{\underset{|}{\overset{|}{C}}}})_n$

4. Label the following reactions as either addition or condensation polymerizations.

(1) 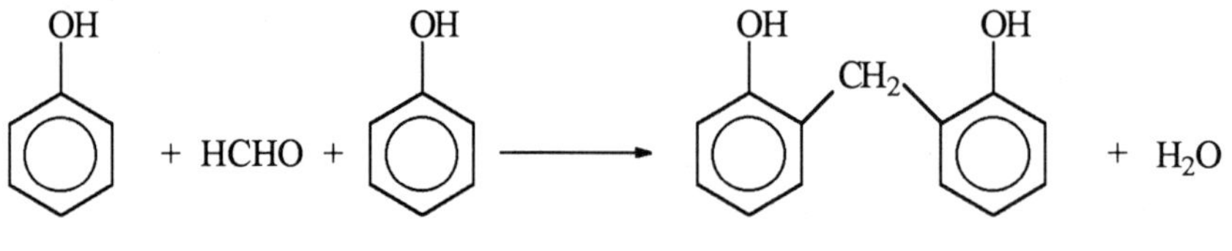 $\underset{}{\bigcirc}^{OH} + HCHO + \underset{}{\bigcirc}^{OH} \longrightarrow \underset{}{\bigcirc}^{OH}-CH_2-\underset{}{\bigcirc}^{OH} + H_2O$

(2) 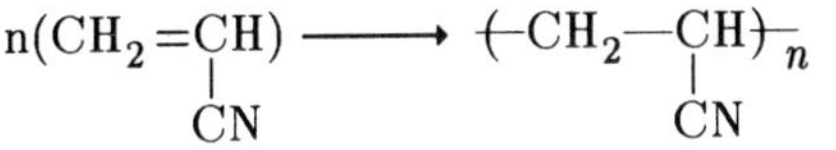 $n(CH_2=CH) \longrightarrow -(CH_2-CH)_n-$
       $|$ $|$
       CN CN

---

(3) 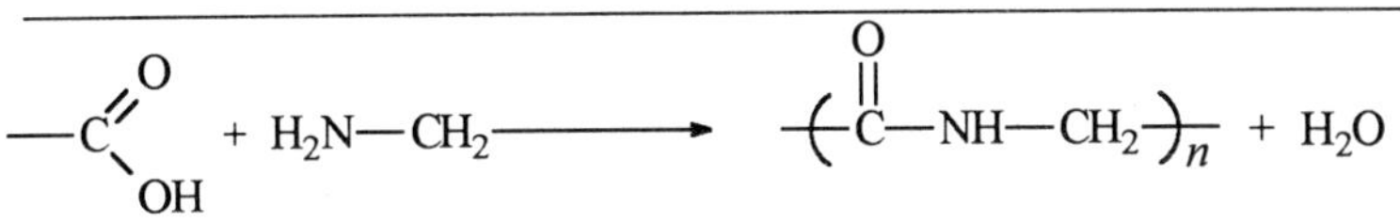 $-C\overset{O}{\underset{OH}{\diagdown}} + H_2N-CH_2- \longrightarrow -(\overset{O}{\overset{\|}{C}}-NH-CH_2)_n- + H_2O$

---

(4) 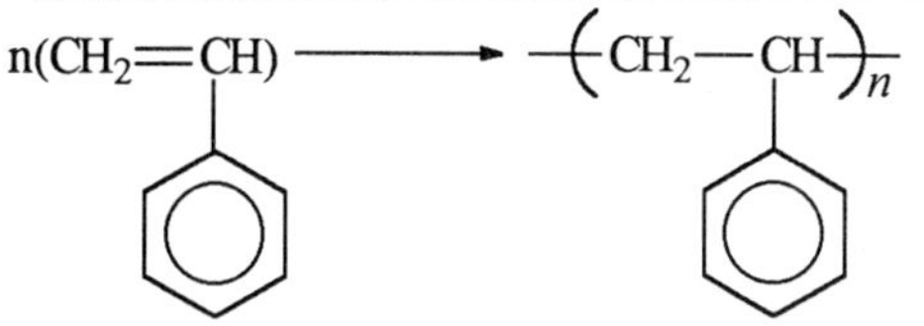 $n(CH_2=CH) \longrightarrow -(CH_2-CH)_n-$

   (with phenyl group on the CH)

---

(5) 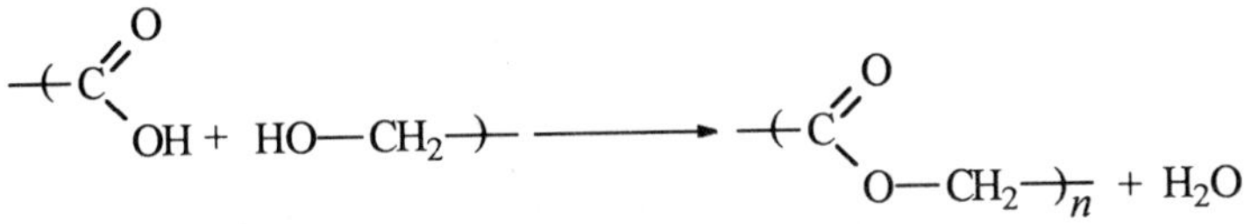 $-(C\overset{O}{\underset{OH}{\diagdown}} + HO-CH_2)- \longrightarrow -(C\overset{O}{\underset{O-CH_2}{\diagdown}})_n- + H_2O$

---

5. What is the difference between a thermoplastic and a thermosetting polymer? How does the difference in structure lead to their difference in softening ability?

6. Both proteins and nucleic acids are condensation polymers. What type of functional group provides the linkage between protein monomers? between nucleic acid monomers?

7. Would the following monomer most likely form a thermoplastic or a thermosetting polymer? Why?

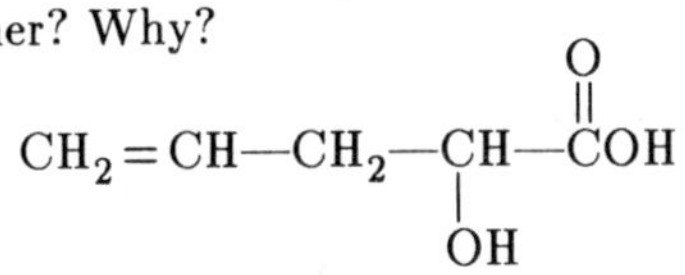

$$CH_2=CH-CH_2-\underset{OH}{CH}-\overset{O}{\overset{\|}{C}}OH$$

8. Identify which of the following structures is the *cis* configuration for natural rubber and which is the *trans* configuration.

(1)

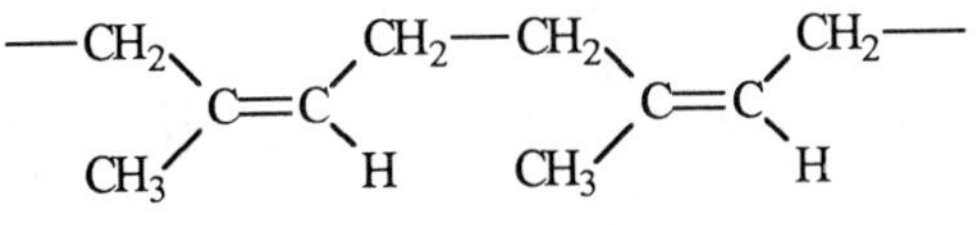

$$-CH_2\diagdown \diagup CH_2-CH_2\diagdown \diagup CH_2-$$
$$C=C C=C$$
$$\diagup CH_3 \diagdown H \diagup CH_3 \diagdown H$$

− 253 −

(2)

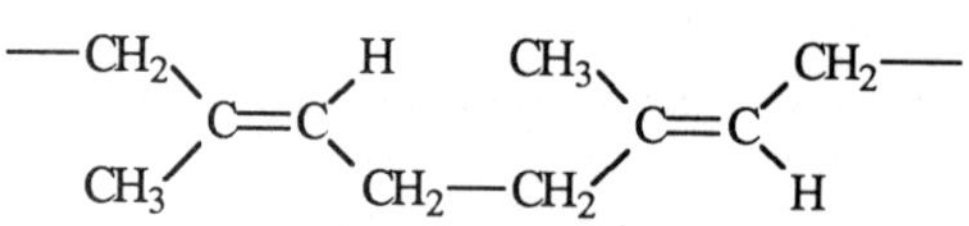

$-CH_2HCH_3CH_2-$
$C=CC=C$
$CH_3CH_2-CH_2H$

9. Would you expect the following molecules to polymerize via addition or condensation? Explain your choice?

   (a) 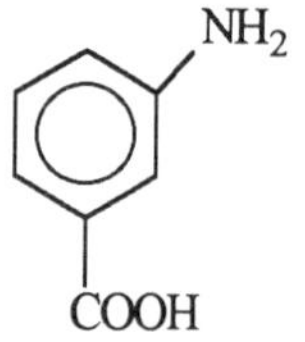 m-aminobenzoic acid (NH$_2$ and COOH on benzene ring)

   (b) $CH_3-C\equiv C-CH_2CH_3$

   (c) $HOCH_2CH_2CH_2OH$

10. What is the by-product split out during the following condensation reactions? Draw the expected polymer.

    (a) 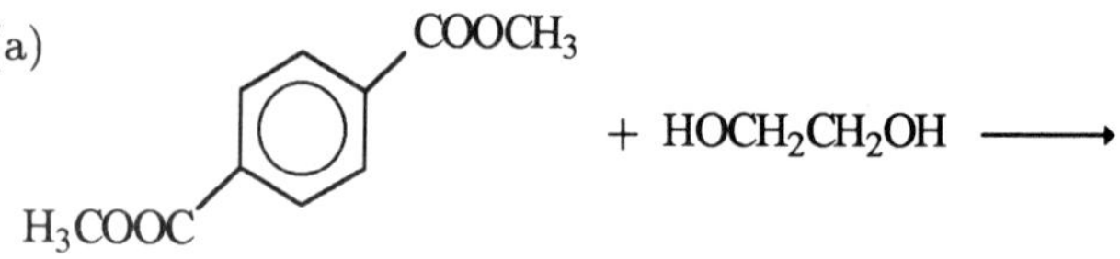 dimethyl terephthalate + $HOCH_2CH_2OH \longrightarrow$

    (b) 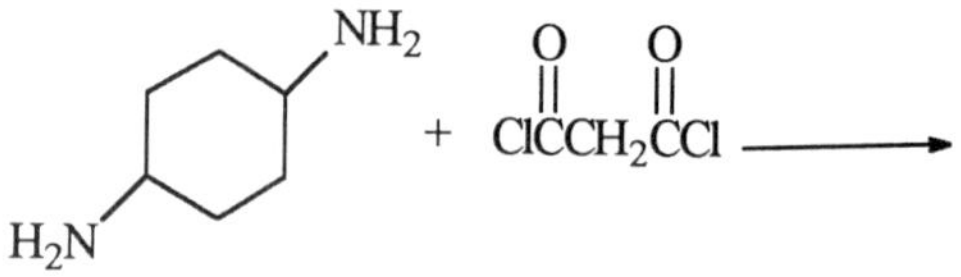 1,4-diaminocyclohexane + $ClCCH_2CCl$ (with two C=O) $\longrightarrow$

11. What is the structure of the condensation polymer formed from urea and formaldehyde?

12. What is perhaps the biggest obstacle to recycling plastic and what is being done about it?

13. You are given an unlimited supply of $HOCH_2CH_2CH_2OH$ and asked to make the following polymer. Propose a synthesis, indicating what type of reaction is occurring at each step.

$$-\overset{O}{\underset{\|}{C}}-CH_2\overset{O}{\underset{\|}{C}}-O-CH_2CH_2CH_2-O-\overset{O}{\underset{\|}{C}}-CH_2-\overset{O}{\underset{\|}{C}}-O-CH_2CH_2CH_2-O-$$

– 254 –

RECAP SECTION

Chapter 26 deals with the topic of polymers, or plastics, as we commonly call them. It has been proposed that historians term the twentieth century the "Age of Plastics" in the same manner that earlier ages are called the Stone Age and the Iron Age. The different types of plastics are explored, their sources, synthesis, and uses. Plastics have a tremendous and pervasive influence on modern life. They are involved in everything we do from birth to death. Increasingly, however, we are faced with the problem of recycling.

ANSWERS TO QUESTIONS

1. (1) polymerization (2) polymer (3) monomer
   (4) stryene (5) macromolecule (6) Bakelite
   (7) thermoplastic (8) thermosetting (9) cross-linked
   (10) addition (11) condensation (12) addition
   (13) condensation (14) Charles Goodyear (15) sulfur
   (16) cross-link

2. (1) chain propagating (2) chain terminating (3) chain initiating
   (4) chain propagating (5) chain terminating order: (3), (1), (4), (2) or (5)

3. (1) c (2) g (3) f
   (4) d (5) b (6) a
   (7) e (8) f monomer for h: $CH_2=C(CH_3)_2$

4. (1) condensation (2) addition (3) condensation
   (4) addition (5) condensation

5. Thermoplastic polymers soften on reheating while thermosetting polymers do not. Thermoplastic polymers are generally polymers which have minimal cross-linking between chains which are therfore held together mainly by weak intermolecular forces. These forces can be easily overcome by raising the temperature thus allowing the chains to alter the distances between them (making the polymer soft). Thermosetting polymers are highly cross-linked and form a network structure that is stable to change in temperature.

6. Proteins are polymers of amino acids which contain carboxylic acid and amine functional groups. The linkage is an amide bond. Nucleic acids are polymers of sugar molecules and phosphate. The linkage of nucleic acid monomers is a phosphate diester.

7. The molecule has three functional groups which can be used for polymerization reactions and should therefore be able to crosslink forming a thermosetting polymer.

8. (1) is *cis* form (2) is *trans* form

9. Addition polymers generally have carbon-carbon multiple bonds (e.g., double or triple bonds), whereas con-densation polymers must be at least bifunctional.

   (a) condensation (b) addition (c) condensation

10. (a)

$$\begin{array}{c}\left(\mathrm{C}(\text{benzene})\mathrm{C}-\mathrm{O}-\mathrm{CH}_2-\mathrm{CH}_2-\mathrm{O}-\mathrm{C}-\text{benzene}-\mathrm{C}-\mathrm{O}-\mathrm{CH}_2-\mathrm{CH}_2-\mathrm{O}\right)_n\end{array}$$

CH_3OH is split out

(b) $\left(-NH-\bigcirc-NHCCH_2C-NH-\bigcirc-NHCCH_2C-\right)_n$ HCl is split out

11. $n\,H\overset{O}{\underset{\|}{C}}H + n\,H_2N\overset{O}{\underset{\|}{C}}NH_2 \xrightarrow{-H_2O} \left(CH_2NH\overset{O}{\underset{\|}{C}}NHCH_2NH\overset{O}{\underset{\|}{C}}NH\right)_n$

12. Thermoplastics are easily recycled because they can be melted and reformed easily. However, the many different types of plastics poses the problem of efficiently separating them. Labelling the plastics with a number helps identify which polymers were used to make the particular item.

13. The monomers are $HO\overset{O}{\underset{\|}{C}}-CH_2-\overset{O}{\underset{\|}{C}}OH$ and $HOCH_2CH_2CH_2OH$. The carboxylic acid can be synthesized from the alcohol by oxidizing the primary diol. A condensation (esterification) reaction of the dicarboxylic acid with the diol should produce the polymer.

$HOCH_2CH_2CH_2OH \xrightarrow{[O]} HO\overset{O}{\underset{\|}{C}}CH_2\overset{O}{\underset{\|}{C}}OH$ (oxidation)

$HO\overset{O}{\underset{\|}{C}}CH_2\overset{O}{\underset{\|}{C}}OH + HOCH_2CH_2CH_2OH \longrightarrow$ polymer   (condensation)

CHAPTER TWENTY-SEVEN

Stereoisomerism

SELF-EVALUATION SECTION

1. Examine the following molecules. Determine if there are asymmetric carbon atoms present and indicate their location.

   (1) 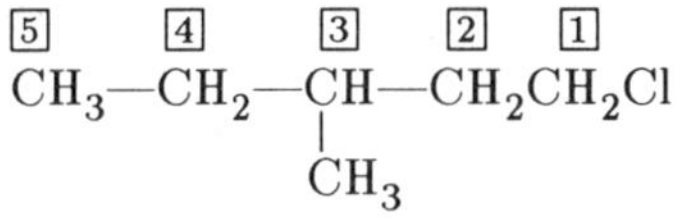 $\overset{[5]}{CH_3}-\overset{[4]}{CH_2}-\overset{[3]}{\underset{\underset{CH_3}{|}}{CH}}-\overset{[2]}{CH_2}\overset{[1]}{CH_2}Cl$ _____

   (2) $\overset{[4]}{CH_3}-\overset{[3]}{\underset{\underset{CH_3}{|}}{\overset{\overset{CH_3}{|}}{C}}}-\overset{[2]}{CH_2}\overset{[1]}{CH_3}$ _____

   (3) $\overset{[3]}{CH_3}-\overset{[2]}{CH_2}-\overset{[1]}{C}\begin{matrix}\nearrow O \\ \searrow OH\end{matrix}$ _____

   (4) $\overset{[4]}{CH_3}-\overset{[3]}{CHCl}-\overset{[2]}{CH_2}-\overset{[1]}{CH_3}$ _____

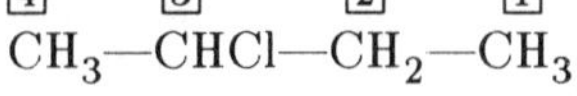

   (5) 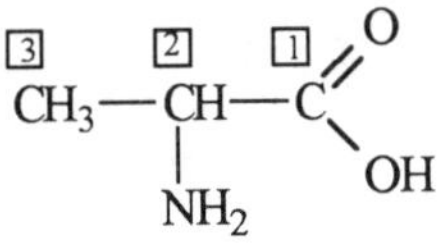 $\overset{[3]}{CH_3}-\overset{[2]}{\underset{\underset{NH_2}{|}}{CH}}-\overset{[1]}{C}\begin{matrix}\nearrow O \\ \searrow OH\end{matrix}$ _____

   (6) 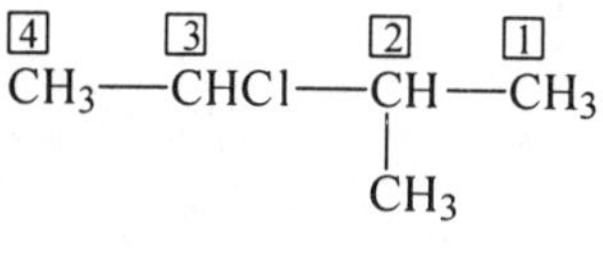 $\overset{[4]}{CH_3}-\overset{[3]}{CHCl}-\overset{[2]}{\underset{\underset{CH_3}{|}}{CH}}-\overset{[1]}{CH_3}$ _____

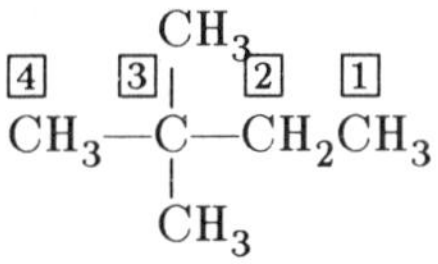

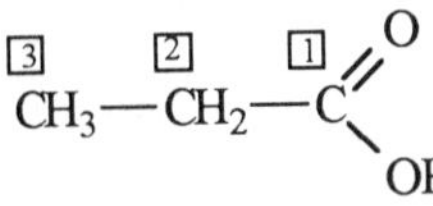

— 257 —

(7)   ③   ②   ①
     CH₂Cl—CHCl—CHClBr   _____

(8)
```
        O
    H—C⃰
     ②│ ①
    H—C—OH
     ③│
    H—C—OH
     ④│
   HO—C—H
     ⑤│
      CH₂OH
```
                                          _____

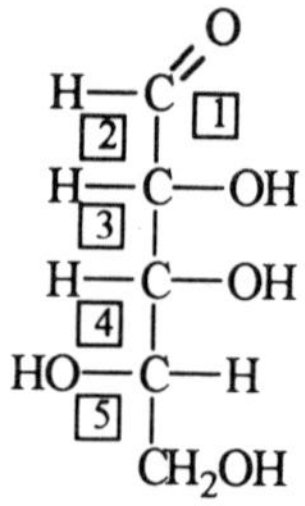

2. Examine the pairs of molecules. Indicate whether the pairs are enantiomers, disastereomers, or neither.

(1) 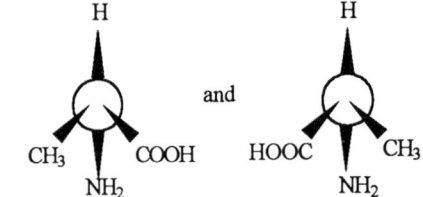   _____

(2) 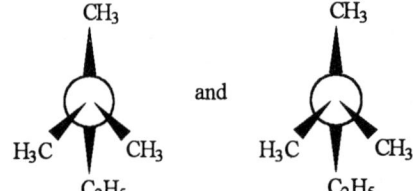   _____

(3)   CH₃   CH₃
       │   │
    H—C—Br   and   H—C—Br
       │   │
    H—C—Cl   Cl—C—H   _____
       │   │
      CH₃   CH₃

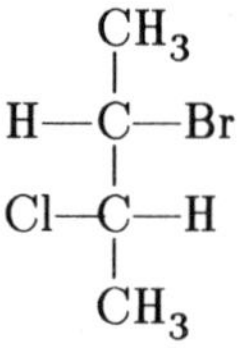

(4)   CH₃   CH₃
       │   │
    H—C—Br   and   Br—C—H
       │   │
    Cl—C—H   H—C—Cl   _____
       │   │
      CH₃   CH₃

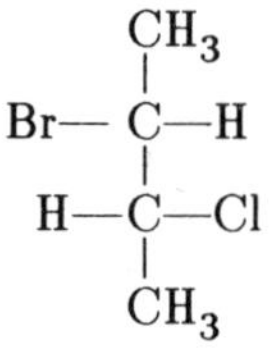

(5)   CH₃   CH₃
    Br─┼─H   H─┼─Cl
    Cl─┼─H   and   H─┼─Br   _____
       │   │
      CH₃   CH₃

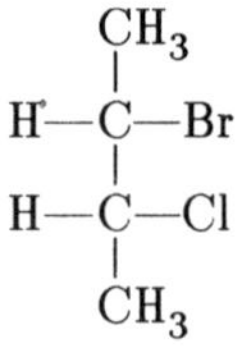

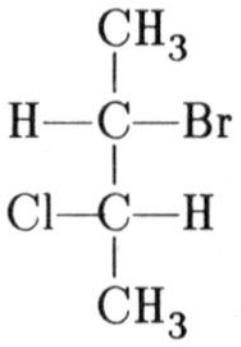

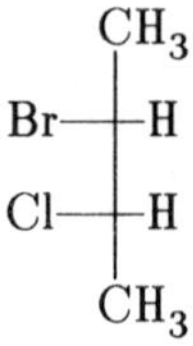

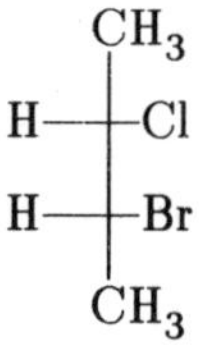

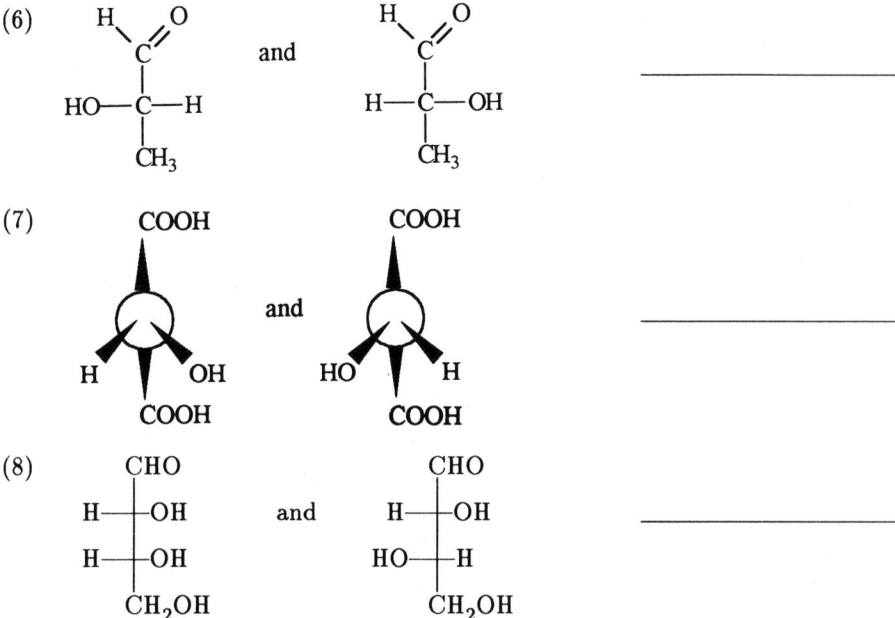

3. Fill in the blank space or circle the appropriate response.

When polarized light is passed through solutions of certain organic compounds, the plane of light is rotated. Such compounds are said to be (1) _____ active. If the plane of light is rotated to the right, the compound is said to be (2) _____ , and if the rotation is to the left, the compound is termed (3) _____ . The famous French scientist Louis (4) _____ is credited with discovering that crystals of an organic salt, sodium ammonium tartrate, existed in two distinct forms. When he separated the two forms he found that either kind of crystal would rotate plane polarized light, but in opposite directions.

In 1874 it was concluded that in the molecule of an optically active substance the key factor is the presence of a(an) (5) <u>asymmetric/dissimilar</u> carbon atom. Such a carbon atom has (6) <u>three/four</u> different (7) _____ attached to it.

Optically active isomers and (8) _____ isomers are examples of stereoisomers.

When two stereoisomers are related to each other as your left hand is to your right hand, the isomers are (9) _____ images of each other. The technical term is (10) <u>enantiomer/optimer</u>.

Equal amounts of mirror image isomers form mixtures called (11) _____ mixtures. During the biological synthesis of potentially optically active compounds, it has been found that racemic mixtures (12) <u>are/are not</u> formed. This result suggests that biological catalysts or (13) _____ are stereospecific.

The physical properties of enantiomers are (14) alike/different except for optical rotation. On the other hand, if a racemic mixture is present, there is (15) no/ some optical rotation even though melting points, solubilities, and specific gravity of the mixture are (16) strikingly different/very similar to the same properties in the isolated enantiomers.

4. Which of the following projection formulas represent (a) enantiomers (b) identical molecules?

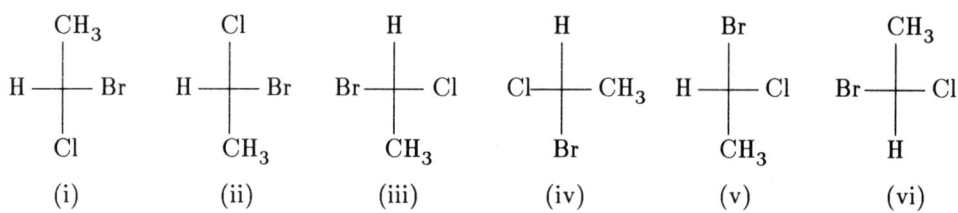

5. Examine the following structural formulas. The open circle represents the possible location of an asymmetric carbon atom. Which of the formulas show an optically active compound?

6. Examine the following molecule. Determine whether it is chiral and if so how many asymmetric carbon atoms there are. Then draw the mirror image pairs and determine if they are superimposable. Label the enantiomers, diastereomers, and meso forms, if any.

   $CH_2Cl-CHCl-CHClBr$

7. Using the information in question 1, calculate the maximum number of stereoisomers possible for each chiral molecule.

8. How many optical isomers exist for a molecule having the following structural formula? Draw them.

   $CH_3CH_2CHClCHClCH_2CH_3$

9. Can either of the following compounds exist as meso forms? Draw projection formulas to be sure of your answer.

   (1) 2-bromobutane   (2) 2,4-dibromopentane

10. What is the maximum possible number of stereoisomers for the following molecule? How many of them would be enantiomers of the molecule below? What is the name given the relationship of the other stereoisomers to the molecule below?

11. Jennifer ran a reaction which she knew should produce molecules with at least one chiral center. However, the specific rotation of a solution of her reaction products was zero. What possible explanations do you propose?

12. Examine the following two molecules. Is either one a chiral molecule? If so, draw a projection formula of the enantiomers.

    (1) $CH_3CHClCH_2CH_3$ (2) CH_3CH_2COOH

13. Find at least one example of each of the following:
    (a) pair of enantiomers
    (b) pair of diastereomers
    (c) meso compound

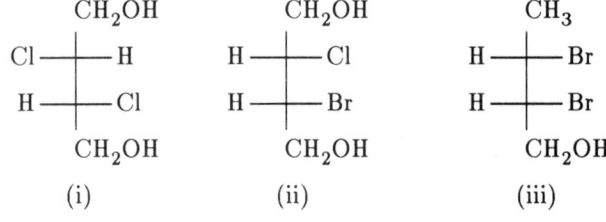

```
      CH₂OH              CH₂OH              CH₃
   H ─┼─ Cl          Cl ─┼─ H           HO ─┼─ H
   Br ─┼─ H           H ─┼─ Cl          HO ─┼─ H
      CH₂OH               CH₃               CH₃
       (iv)                (v)               (vi)
```

RECAP SECTION

Chapter 27 has presented a topic of unique importance to organic chemistry—stereoisomerism. The insights an understanding of this subject affords the scientist are numerous. As an example, recall that living systems synthesize and utilize only one enantiomer if an asymmetric carbon atom exists in a molecule. Therefore, various drugs that are manufactured as racemic mixtures are only utilized to half their total concentration by living systems. Only one enantiomer of a particular amino acid is found to be incorporated into proteins. The introductory material in the chapter, even though it is brief, has indicated some of the complexities involved with studying stereoisomers.

ANSWERS TO QUESTIONS

1. An asymmetric carbon has four different groups attached.
   (1) Carbon number 3 is asymmetric. Carbon number 1 is not, because there are two hydrogens attached.
   (2) There are no asymmetric carbon atoms.
   (3) There are no asymmetric carbon atoms.
   (4) Carbon number 3 is asymmetric.
   (5) Carbon number 2 is asymmetric.
   (6) Carbon number 3 is asymmetric. Carbon number 2 is not, because it has two methyl groups attached.
   (7) Carbon number 1 and carbon number 2 are asymmetric.
   (8) Carbons number 2, 3, and 4 are asymmetric.

2. (1) enantiomers
   (2) neither (there are 3 CH₃ groups attached to the central carbon atom, hence no asymmetry)
   (3) diastereomers
   (4) enantiomers
   (5) neither; they are identical molecules (rotate 180° to verify)
   (6) enantiomers
   (7) neither; there are two identical COOH groups
   (8) diastereomers

3. (1) optically   (2) dextrorotatory (3) levorotatory   (4) Pasteur
   (5) asymmetric   (6) four   (7) atoms or groups   (8) geometric
   (9) mirror   (10) enantiomer   (11) racemic   (12) are not
  (13) enzymes   (14) alike   (15) no   (16) strikingly different

4. Molecules are identical if they are superimposable. If two groups coincide and the other two do not, the molecules are enantiomers of each other. Manipulations which are allowed for these projection formula are rotation of 180° or sliding the molecule over another. If the relationship is not obvious, count the number of successive group interchanges needed to make the formulas identical. An odd number of interchanges means the molecules are enantiomers.

   (a) enantiomers: (i, ii), (i, iii), (i, iv), (v, ii), (v, iii), (v, iv), (vi, ii), (vi, iii), (vi, iv)

   (b) identical: (i, v, vi) and (ii, iii, iv)

5. (2), (5), and (6) are optically active. The other compounds do not have four different atoms or groups attached to the possible asymmetric carbon atom.

6. The molecule is chiral and has two asymmetric carbon atoms. Mirror images would be as follows:

$$
\begin{array}{c}
CH_2Cl \\
| \\
Cl-C-H \\
| \\
Cl-C-H \\
| \\
Br
\end{array}
\qquad
\begin{array}{c}
CH_2Cl \\
| \\
H-C-Cl \\
| \\
H-C-Cl \\
| \\
Br
\end{array}
$$

I   II

$$
\begin{array}{c}
CH_2Cl \\
| \\
Cl-C-H \\
| \\
H-C-Cl \\
| \\
Br
\end{array}
\qquad
\begin{array}{c}
CH_2Cl \\
| \\
H-C-Cl \\
| \\
Cl-C-H \\
| \\
Br
\end{array}
$$

III   IV

I and II are a pair of enantiomers as are III and IV. Diastereomer pairs are I and III, I and IV, II and III, and II and IV. There are no meso forms.

7. Using the molecules in question 1, we found that
   (1) has one asymmetric carbon atom. Therefore,
       $2^1 = 2$ stereoisomers
   (4) $2^1 = 2$ stereoisomers
   (5) $2^1 = 2$ stereoisomers
   (6) $2^1 = 2$ stereoisomers
   (7) $2^2 = 4$ stereoisomers
   (8) $2^3 = 8$ stereoisomers

8. This molecule has two chiral carbons and therefore has a maximum of 2^n $(n = 2) = 4$ stereoisomers. However, one of the stereoisomers is identical to another because they have a plane of symmetry.

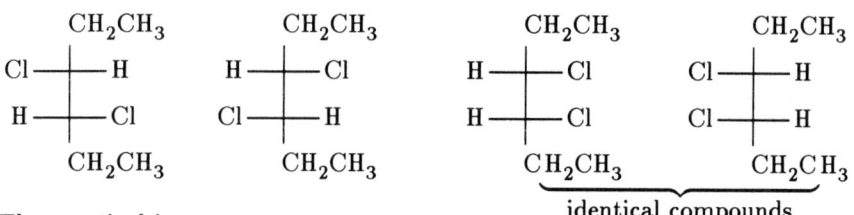

Three optical isomers.   identical compounds

9. (1) 2-bromobutane does not exist as a meso form.
    (2) 2,4-dibromobutane has a plane of symmetry and therefore exists as a meso form.

10. Maximum number of stereoisomers is 2^n. The molecule below has 4 chiral centers (indicated by $*$) and therefore there are 16 possible stereoisomers. Only one could be a mirror-image so the molecule has only one enantiomer. The other stereoisomers are called diastereomers of the molecule below.

11. Jennifer may have a racemic mixture (an equal mixture of a pair of enantiomers) or her molecule may contain chiral carbons but also have a plane of symmetry (a meso compound). Both situations give rise to a solution having no optical rotation.

12. (1) is chiral, (2) is not. The enantiomer projection formulas would be

    mirror images non-superimposable

13. (a) (i) and (v)   (b) (ii) and (iv)   (c) (vi)

CHAPTER TWENTY-EIGHT

Carbohydrates

SELF-EVALUATION SECTION

1. For each of the following carbohydrate molecules, pick out all of the correct descriptive terms from the list that apply to the particular molecule.

   triose
   tetrose
   pentose
   hexose
   number of asymmetric carbon atoms –
       e.g., 1, 2, 3, 4, or 5 atoms
   number of possible optical isomers –
       e.g., 2, 4, 8, 16 or 32 isomers

   reducing sugar
   nonreducing sugar
   D configuration
   L configuration
   aldose
   ketose

   (1)
       H   O
         C
     H—————OH
        CH$_2$OH

   (2)
       CH$_2$OH
     H—————OH
       =O
     H—————OH
       CH$_2$OH

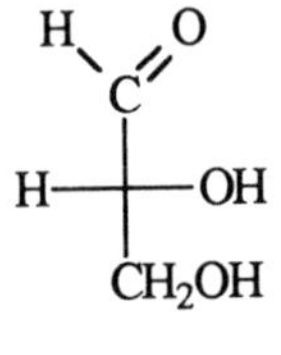

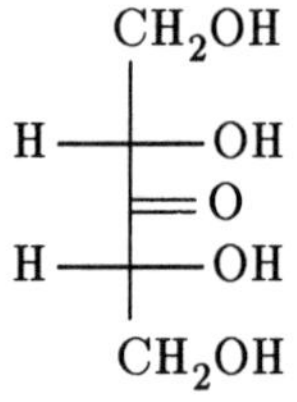

(3)
```
    H   O
     \\//
      C
  H—|—OH
 HO—|—H
  H—|—OH
  H—|—OH
     CH₂OH
```

(4)
```
     CH₂OH
      |
      C=O
 HO—|—H
  H—|—H
  H—|—OH
     CH₂OH
```

(5)
```
    H   O
     \\//
      C
  H—|—OH
  H—|—OH
 HO—|—H
     CH₂OH
```

(6)
```
    H   O
     \\//
      C
  H—|—OH
  H—|—OH
  H—|—OH
     CH₂OH
```

(7)
```
     CH₂OH
      |
      C=O
 HO—|—H
     CH₂OH
```

(8)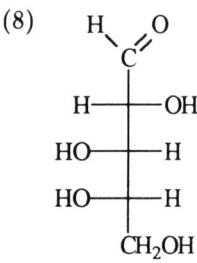

2. Fill in the blank space or circle the appropriate response.

Carbohydrates, one of the three principal classes of foods, contain only three elements: (1) _____, (2) _____, and (3) _____ .

The name "carbohydrate" is derived from the French *hydrates de carbone* because the empirical formula in many cases is (4) $C \cdot HO/(C \cdot H_2O)_n$. Carbohydrates are actually aldehydes or ketones that contain many (5) hydroxy/aromatic groups. The simplest carbohydrates are glyceraldehydes and dihydroxy acetone.

A classification system for carbohydrates is based on the number of units linked together. If there is only one unit, the carbohydrate is termed a (6) _____ . Two units linked together are a (7) _____ , such as sucrose, while long chain carbohydrates, such as glycogen and cellulose, are (8) _____ .

The most common monosaccharide, (9) _____ , has (10) four/six asymmetric carbon atoms. Thus, the number of isomers related to glucose is (11) _____ . All of these have been synthesized, but only D-glucose and D-galactose are of biological importance.

Another common hexose, fructose, is a(an) (12) aldose/ketose, and when linked to glucose, it forms the disaccharide (13) _____ . Other disaccharides are (14) _____ and lactose, which is known also as (15) _____ .

All monosaccharides and most disaccharides are reducing sugars capable of reducing silver ions and copper ions. The only common sugar that is not a reducing sugar is the disaccharide (16) _____ .

3. Draw the possible isomers of a D-aldose that contains four carbons (tetrose): $C_4H_8O_4$ is the formula. Use Fischer projection formulas.

4. (a) What is the difference between a diastereomer and an epimer?
   (b) Are all diastereomers epimers or all epimers diastereomers or neither?
   (c) How are the following structures related? (Be as specific as possible.)

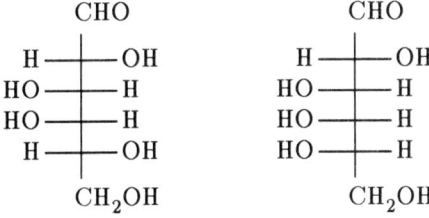

5. Examine the carbohydrate formulas below and then answer the questions.

   (1) Identify the pyranose and furanose ring structures.
   (2) Identify the Fischer projection and Haworth formulas.
   (3) Identify the hemiacetal forms.
   (4) Identify (d) as either D or L.

6. Match the disaccharides on the left with their two constituent monosaccharides from the list on the right.

Disaccharides	Monosaccharides
(1) sucrose	glucose
(2) lactose	fructose
(3) maltose	galactose

7. Match a possible source from the right with the sugar listed on the left.

Sugar	Source
(1) glucose	fruit juice
(2) galactose	sugar beets
(3) fructose	starch
(4) sucrose	sucrose
(5) lactose	lactose
(6) maltose	pectin
	honey
	milk
	sugar cane
	sprouting grain

8. Match the polysaccharide with the correct descriptive statement.

(1)	starch	a.	glucose units, α-1,4 linkage, highly branched
(2)	cellulose	b.	fructose units, 1,2-linkage
(3)	glycogen	c.	glucose units, α-1,4 linkage, moderately branched
(4)	inulin	d.	glucose units, β-1,4 linkage

9. Fill in the blank space or circle the appropriate response.

   The hemiacetal structure of carbohydrates involves an alcohol linkage and a(n) (1) _____ linkage on the same carbon atom. Acetal structures have (2) _____ ether linkages to the same carbon atom. The hemiacetal or cyclic form is unstable and hydrolyzes easily to the open form, which can reform to the other diastereomer. The symbols used for the two forms are alpha (α) and beta (3) _____. Each of the diastereomers rotates polarized light and in solution there is an equilibrium between the forms. Diastereomers that differ in spatial arrangement only on C-1 carbon are called (4) _____. The switching back and forth from α to β forms is called (5) _____ and occurs in hemiacetals but not (6) _____. Glycosides such as cellulose and starch are only ether linkages and are therefore (7) hemiacetal/acetal structures and (8) do/do not exhibit mutarotation. Lactose and maltose undergo mutarotation, which means at least one of their monosaccharides has a (9) _____ ring structure. Sucrose does not undergo mutarotation, which means only (10) _____ linkages.

10. Fully describe all the information given in the name of the following molecule. Use that information to draw the molecule (+)-D-α-glucopyranose.

11. Describe a possible glycosidic linkage of a disaccharide made up of D-glucose and D-galactose which would yield a *non*-reducing sugar.

12. Monosaccharides can undergo several chemical reactions. Use the following reaction conditions to answer the questions below.

    Br_2/H_2O warm HNO_3 $C_6H_5NHNH_2$ H_2/Pt

    (a) Which reagent(s) will produce a meso compound when used with D-galactose?
    (b) Which reagent will produce D-galactonic acid from D-glucose?
    (c) Which reagent(s) affects more than one carbon of an aldohexose?
    (d) Which reagent produces identical products from D-glucose, D-mannose, and D-fructose?

13. Complete the following reactions.

    (1) Benedict's test. What is visual evidence of a positive test?

    $$R-\overset{H}{\underset{\|}{C}}=O + 2\,Cu^{2+} \longrightarrow \underline{\qquad} + \underline{\qquad} + 3\,H_2O$$
    (blue)

(2) Oxidation. For the same tetrose, give products formed under two different oxidizing conditions.

$$\begin{array}{c} H\diagdown\;\;\diagup O \\ C \\ | \\ HO-C-H \\ | \\ H-C-OH \\ | \\ CH_2OH \end{array} \quad \begin{array}{c} Br_2 + H_2O \nearrow \\ \\ \searrow \text{warm} \\ HNO_3 \end{array}$$

(3) Reduction of D-glucose with H_2/Pt

$$\begin{array}{c} H\diagdown\;\;\diagup O \\ C \\ | \\ H-C-OH \\ | \\ HO-C-H \\ | \\ H-C-OH \\ | \\ H-C-OH \\ | \\ CH_2OH \end{array} \xrightarrow{\;H_2\;}_{Pt}$$

14. What are four differences between cellulose and starch? Do they have anything in common? Are they addition or condensation polymers?

15. Using Table 28.1, indicate which functional group of the pairs given is higher in relative potential chemical energy.

   (1) CH_3 group versus primary alcohol
   (2) aldehyde versus methylene carbon
   (3) secondary alcohol versus ketone

16. Determine the oxidation states of each carbon atom in the first four carbohydrate molecules of Question 1. Number the uppermost carbon atom in each molecule as 1, the second as 2 and so on. Which carbon atom in each molecule may provide the most energy for a biological system?

17. Examine the Fischer projection formulas shown for several pentoses. Pick out the formulas that will give identical ozazones following reaction with phenyl hydrazine.

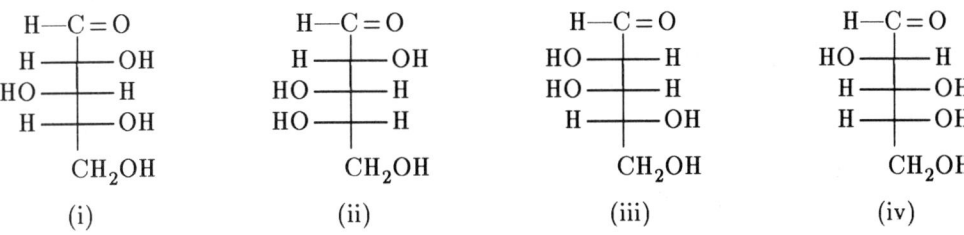

18. Why are sugar substitutes so sought after?

19. What structural features make carbohydrates so important?

RECAP SECTION

Chapter 28 has presented the structural information about one of three important classes of organic compounds utilized as foodstuffs—carbohydrates, fats, and proteins. We have learned the differences between monosaccharides, disaccharides, and polysaccharides, and the extensive number of monosaccharide isomers to be found because of asymmetric carbon atoms. Methods of classification and a brief introduction to the everyday uses and problems associated with the usage of these compounds are mentioned. A more in-depth look at the biological importance of this class of compounds will be presented in a later chapter.

ANSWERS TO QUESTIONS

1. (1) triose, 1 asymmetric carbon atom, 2 isomers, reducing sugar, D configuration, aldose
   (2) pentose, 2 asymmetric carbon atoms, 4 isomers, nonreducing sugar, D configuration, ketose
   (3) hexose, 4 asymmetric carbon atoms, 16 isomers, reducing sugar, D configuration, aldose
   (4) hexose, 2 asymmetric carbon atoms, 4 isomers, reducing sugar, D configuration, ketose
   (5) pentose, 3 asymmetric carbon atoms, 8 isomers, reducing sugar, L configuration, aldose
   (6) pentose, 3 asymmetric carbon atoms, 8 isomers, reducing sugar, D configuration, aldose
   (7) tetrose, 1 asymmetric carbon atom, 2 isomers, reducing sugar, L configuration, ketose
   (8) pentose, 3 asymmetric carbon atoms, 8 isomers, reducing sugar, L configuration, aldose

2. (1) carbon (2) oxygen (3) hydrogen (in any order)
   (4) $(C \cdot H_2O)_n$ (5) hydroxy (6) monosaccharide
   (7) disaccharide (8) polysaccharides (9) glucose (10) four
   (11) 16 (12) ketose (13) sucrose (14) maltose
   (15) milk sugar (16) sucrose

3. Since we are dealing with a tetrose, which is an aldose, we will have two asymmetric carbon atoms at carbon number 2 and 3. We have a D sugar, which means the hydroxy group on carbon number 3 is to the right.

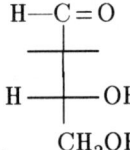

If there are two asymmetric carbon atoms, then there are 2^2 or four isomers, two pairs of enantiomers. The two D isomers are, therefore,

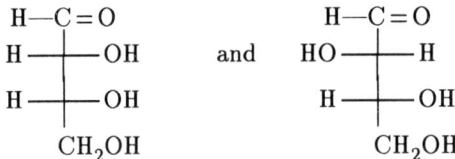

4. (a) Diastereomers are stereoisomers which differ spacially at a minimum of one chiral center but *not* all chiral centers. Epimers are special types of diastereomers. They differ in configuration at only *one* chiral center.
   (b) All epimers are diastereomers.
   (c) These molecules are epimers (carbon 5 is different).

5. (1) a and b are pyranose, c is furanose.
   (2) Fischer projections are a and d, and Haworth formulas are b and c.
   (3) Hemiacetal forms are a, b, c.
   (4) d is "D".

6. (1) sucrose—glucose and fructose
   (2) lactose—glucose and galactose
   (3) maltose—two glucose units

7. (1) glucose—starch, sucrose, lactose, and maltose
   (2) galactose—lactose, pectin
   (3) fructose—honey, sucrose, fruit juice
   (4) sucrose—sugar cane, sugar beets
   (5) lactose—milk
   (6) maltose—sprouting grain

8. (1) starch—c (2) cellulose—d
   (3) glycogen—a (4) inulin—b

9. (1) ether (2) two (3) β (4) anomers
   (5) mutarotation (6) acetals (7) acetal (8) do not
   (9) hemiacetal (10) acetal or ether

10. (+)—means a solution of the sugar rotates plane-polarized light to the right (dextrorotatory).

    D—identifies which isomer of the sugar (the penultimate hydroxyl group is on the right in the Fischer projection).

    α—indicates which anomer was formed when the molecule cyclized to the hemiacetal form. The new stereocenter has the hydroxyl group in the axial position (down).

    gluco—identifies the sugar (glucose) and therefore the arrangement of the hydroxyl groups on the chiral carbons.

    pyranose—indicates that when the molecule cyclized, a six-membered ring was formed.

    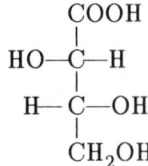

    (+)-D-α-glucopyranose

11. Non-reducing sugars do not have free hydroxyl groups (OH) attached to their anomeric carbons. Therefore a (1,1) glycosidic bond between D-glucose and D-galactose should yield a non-reducing disaccharide.

12. (a) If carbons 1 and 6 of D-galactose become identical, the molecule is meso. Therefore warm HNO_3, or H_2/Pt will react with D-galactose to produce a meso compound.
    (b) Br_2/H_2O
    (c) Warm HNO_3 oxidizes carbons 1 and 6; phenylhydrazine reacts with carbons 1 and 2.
    (d) phenylhydrazine

13. (1) Products would be $R-C\begin{smallmatrix}O\\O^-\end{smallmatrix}$ and Cu_2O

    The Cu_2O will be a brick-red precipitate.

    (2) Product from $Br_2 + H_2O$ oxidation would be

    $$\begin{array}{c} COOH \\ | \\ HO-C-H \\ | \\ H-C-OH \\ | \\ CH_2OH \end{array}$$

Product from warm HNO_3 oxidation would be

$$\begin{array}{c} COOH \\ | \\ HO-C-H \\ | \\ H-C-OH \\ | \\ COOH \end{array}$$

(3) Reduction of the glucose product

$$\begin{array}{c} CH_2OH \\ | \\ H-C-OH \\ | \\ HO-C-H \\ | \\ H-C-OH \\ | \\ H-C-OH \\ | \\ CH_2OH \end{array}$$

14. Cellulose and starch differ in their shapes, the glycosidic linkage, their solubility in water, and their natural source. They are both made from D-glucose. They are both condensation polymers.

15. (1) CH_3 group (2) methylene carbon (3) secondary alcohol

16. (1) $C_1 = +1$, $C_2 = 0$, $C_3 = -1$.
    C_3 is most reduced and should provide most energy.
    (2) $C_1 = -1$, $C_2 = 0$, $C_3 = +2$, $C_4 = 0$, $C_5 = -1$.
    C_1 and C_5 should provide the most energy.
    (3) $C_1 = +1$, $C_2 = 0$, $C_3 = 0$, $C_4 = 0$, $C_5 = 0$, $C_6 = -1$.
    C_6 should provide the most energy.
    (4) $C_1 = -1$, $C_2 = +2$, $C_3 = 0$, $C_4 = -2$, $C_5 = 0$, $C_6 = -1$.
    C_4 should provide the most energy.

17. (i) and (iii) would give identical ozazones

18. Too much sugar in the diet can lead to several health problems. The high calorie count may cause weight problems, the ease of metabolism by bacteria results in dental problems, and diabetics often cannot handle the sharp rise in blood sugar following a high sugar meal.

    Sugar substitutes—molecules which are either not metabolized or have higher sweetening ability—are seen as reducing health risks caused by too much sugar in the diet.

19. Carbohydrates contain carbonyl (aldehyde or ketone) groups and hydroxyl groups. These functional groups enable carbohydrates to be relatively easy to oxidize thus yielding energy. The polarity of these groups allow many smaller carbohydrates to be water soluble allowing them to travel in the body easily. Polymers of the carbohydrates form strong hydrogen bonds within and between chains thus rendering them insoluble in water and strong enough to provide structural support.

CHAPTER TWENTY-NINE

Lipids

SELF-EVALUATION SECTION

1. Fill in the blank space or circle the appropriate response.
   Lipids are a group of organic compounds classified on the basis of solubility in fat solvents such as (1) _____ . The simple lipids are the fats, oils, and waxes.

   Fats and oils are fatty acid esters of the trihydroxy alcohol (2) _____ . The long-chain R groups of a fat are usually (3) the same/different. The main physical difference between fats and oils is that fats are (4) _____ at room temperature, while oils are (5) _____ . In addition, fats usually come from (6) vegetable/animal sources, while oils are from (7) _____ sources. The difference in physical properties in part can be attributed to the amount of unsaturation in the fatty acid groups. Oils generally have a (8) greater/lesser amount of unsaturation as compared with fats. Oils are hardened by a process of hydrogenation to make them solid. During the reaction, hydrogen is (9) added to/removed from the carbon-carbon double bonds. When compared with carbohydrates and proteins, fats provide a (10) greater/lesser amount of caloric value.

   Waxes are (11) complex/simple esters of (12) long-/short- chain fatty acids and alcohols. They are usually of (13) animal/plant origin and may serve protective functions such as helping reduce water loss from the leaves. Phospholipids, glycolipids, and steroids are examples of (14) simple/complex lipids that serve many regulatory biological functions as vitamins and hormones.

2. Identify the hydrophobic and hydrophilic portions of the following phospholipid.

$$\begin{array}{c} \quad\quad\quad\quad\quad\quad\quad O \\ \quad\quad\quad\quad\quad\quad\quad \| \\ \quad\quad\quad\quad CH_2-O-C-R \\ O \quad\quad | \\ \| \quad\quad | \\ R-C-O-CH \quad\quad O \\ \quad\quad\quad | \quad\quad \| \\ \quad\quad\quad CH_2-O-P-O^- \\ \quad\quad\quad\quad\quad\quad | \\ \quad\quad\quad\quad\quad\quad O_- \end{array}$$

3. How does aspirin affect the conversion of arachidonic acid to prostaglandins?

4. Draw the structure common to all steroids.

5. Determine the oxidation number of the central carbon in the following species and indicate which has the greater potential chemical energy and why.

```
                          OH OH OH
                          |  |  |
   carbohydrate fragment  --- C—C—C ---
                          |  |  |
                          H  H  H


                          H  H  H
                          |  |  |
   fat fragment           --- C—C—C ---
                          |  |  |
                          H  H  H
```

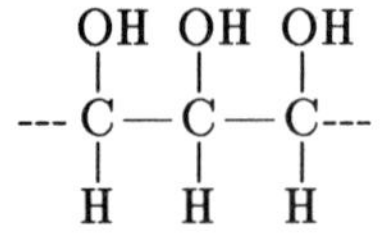

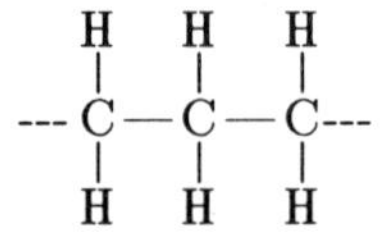

6. Match the phrase of key words from the right hand list with the terms on the left:

   (1) atherosclerosis   a. linoleic, arachidonic, linolenic
   (2) essential fatty acids   b. myelin sheath
   (3) phospholipid   c. converted to ACTH, estrogen, and testosterone
   (4) cholesterol   d. deposition of cholesterol as plaque
   (5) fat   e. thin, semi-permeable cellular barriers
   (6) cephalin   f. lecithins
   (7) sphingomyelin   g. major body reserve of potential energy
   (8) glycolipids   h. essential in blood clotting
   (9) cerebrosides   i. a glucolipid containing complex
   (10) anesthetic   oligosaccharide as their carbohydrate
   (11) ganglioside   j. cell membrane of brain tissue
   (12) sphingosine   k. sphingolipids that contain carbohydrates
   (13) membranes   l. unsaturated amino alcohol
                                m. alters membrane fluidity

7. What do the abbreviations HDL, LDL, VLDL mean? What is the difference in their biological roles? Which is the "good" cholesterol and which is the "bad" cholesterol?

8. What is the difference between a micelle and a liposome found in an aqueous medium?

9. Is the micelle drawn below most likely to be found in an aqueous or organic medium? Why? Draw the micelle you would expect to see in the medium you did not choose?

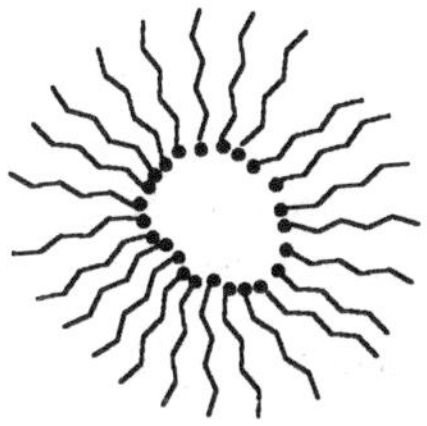

10. Below is a diagram of a typical membrane. Identify each section as being either hydrophobic or hydrophilic.

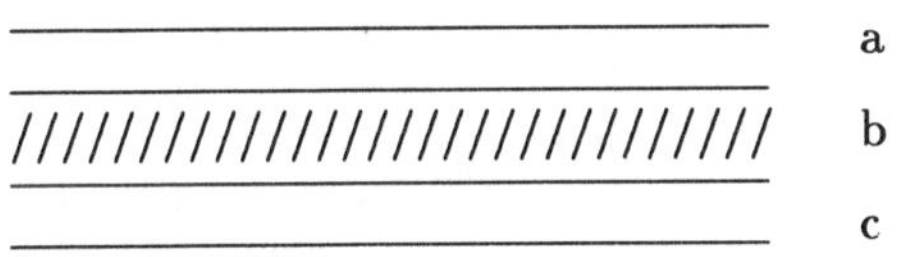

    _____ a
    //////////////////////// b
    _____ c

11. Fill in the blank space or circle the appropriate response.

    The chemical composition of membranes, a lipid (1) bilayer/monolayer, gives rise to questions about how hydrophilic molecules are transported across the membrane from one side to the other. It is now known that specific (2) alcohols/proteins occur within the bilayer and allow for the movement of hydrophilic molecules across the lipid bilayer. If the protein helps the transport without using energy, the process is termed (3) _____ while the movement of molecules from areas of low concentration to high concentration is called (4) _____ and requires energy.

    Nerve cells provide good examples of membrane function. An active transport system, which (5) uses/produces energy, concentrates K^+ ions inside the cell while expelling Na^+ ions. When a neuron transmits a signal, some K^+ ions flow out rapidly and some Na^+ flows in without using energy. This is an example of (6) _____. The rapid ion movement is an electrical nerve impulse.

    The myelin sheath acts as a(an) (7) conductor/insulator around the neuron and when the myelin sheath is damaged or destroyed, nerve transmission is poor resulting in a serious crippling disease, (8) _____.

12. What is the difference between active transport and facilitated diffusion? Which one requires energy and why?

13. What is the difference between an intrinsic and an extrinsic membrane protein?

RECAP SECTION

Chapter 29 presents information about the class of biochemical compounds called lipids. Lipids serve important structural functions in membranes and as potential energy sources for metabolism. Several detailed examples were presented to give you an appreciation of lipid compounds' importance for cells and organisms.

ANSWERS TO QUESTIONS

1. (1) ether, benzene, or carbon tetrachloride
   (2) glycerol
   (3) different
   (4) solid
   (5) liquid
   (6) animal
   (7) vegetable
   (8) greater
   (9) added to
   (10) greater
   (11) simple
   (12) long
   (13) plant
   (14) complex

2. The $-O-\overset{\overset{O}{\|}}{C}-R$ portion is the hydrophobic part

   The $-O-\overset{\overset{O}{\|}}{\underset{\underset{O^-}{|}}{P}}-O^-$ portion is the hydrophilic part

3. Aspirin prevents the enzymatic oxidation of arachidonic acid to prostaglandins.

4.

5. Oxidation number of carbon in the carbohydrate is zero (0) and in the fat the number is –2. Therefore, carbon in a fat is more reduced than in carbohydrate and contains more potential energy. Fats are better sources of energy than carbohydrates because of more reduced carbon.

6. (1) d (2) a (3) f (4) c (5) g (6) h
   (7) b (8) k (9) j (10) m (11) i (12) l (13) e

7. HDL—high density lipoprotein
   LDL—low density protein
   VLDL—very low density lipoprotein

   VLDL—are essentially delivery lipids—they deliver triacylglycerol to fat cells.
   LDL—also delivery lipids—they deliver cholesterol to peripheral tissues.
   HDL—cholesterol scavenger—collects cholesterol and returns it to the liver.

   LDL—considered to be "bad" cholesterol
   HDL—considered to be "good" cholesterol

8. Micelles are usually made up of molecules which contain a polar, hydrophilic head and a nonpolar hydrophobic tail, the latter of which aggregate. The hydrophilic heads are exposed to the aqueous solution. Liposome molecules usually have two hydrophobic chains which result in a different aggregation. Liposomes generally have a water core because they are formed from *two* layers of lipids whose tails face each other.

9. ⌇⌇ ← tail—generally nonpolar tail is hydrophobic. Since the tails are facing outward, this micelle is likely found in an organic medium. The polar heads aggregate away from the nonpolar medium. The structure most likely to exist in an aqueous medium would be

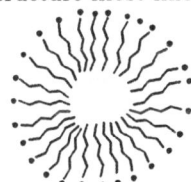

10. membrane (a) hydrophilic (b) hydrophobic (c) hydrophilic

11. (1) bilayer   (2) proteins   (3) facilitated diffusion
    (4) active transport (5) uses   (6) facilitated diffusion
    (7) insulator   (8) MS (multiple sclerosis)

12. Proteins located in the fluid bilayer shuttle molecules through a fluid bilayer into a cell. If energy is required for this transport, the process is called active transport. If the transport does not require energy, the process is called facilitated diffusion.

13. A membrane composed of a lipid bilayer also contains proteins. Proteins found primarily on the surface of the lipid bilayer are termed extrinsic membrane proteins. Intrinsic membrane proteins are found mainly inside the lipid bilayer.

CHAPTER THIRTY

Amino Acids, Polypeptides, and Proteins

SELF-EVALUATION SECTION

1. From the list below, circle the foods high in protein content.

   bread   melon   spaghetti   apples
   fish   chicken   cheese   potatoes
   carrots   nuts   eggs   beans

2. All amino acids contain a carboxylic acid and an amine. Why are only some amino acids classified as acidic or basic?

3. Can neutral amino acids have ionizable side chains?

4. Match the term in column A with the description in Column B.

Column A		Column B
(1) Primary structure	_____	a. Equal number of amino and carboxyl groups
(2) Basic amino acid	_____	b. The type and sequence of amino acids in a protein
(3) Neutral amino acid	_____	c. The higher order of structure found in complex proteins
(4) Tertiary structure	_____	d. More carboxyl groups than amino groups
(5) Acidic amino acid	_____	e. More amino groups than carboxyl groups
(6) Quarternary structure	_____	f. The characteristic shape or conformation of a protein
(7) Secondary structure	_____	g. The pleated or helical structure of a protein

5. Give the product ion when the valine zwitterion reacts with (1) acid and (2) base.

$$\xleftarrow{H^+} \quad \begin{array}{c} CH_3 \\ \diagdown \\ CH-CH-C\diagup\!\!\!\diagup O \\ CH_3\diagup \quad \mid \diagdown O^- \\ ^+NH_3 \end{array} \xrightarrow{OH^-}$$

(1)

$$\begin{array}{c} CH_3 \\ \diagdown \\ CH-CH-C\diagup\!\!\!\diagup O \\ CH_3\diagup \quad \mid \diagdown OH \\ ^+NH_3 \end{array}$$

(2)

$$\begin{array}{c} CH_3 \\ \diagdown \\ CH-CH-C\diagup\!\!\!\diagup O \\ CH_3\diagup \quad \mid \diagdown O^- \\ NH_2 \end{array}$$

6. Fill in the blank space with either the name of the amino acid or its abbreviation.

	Name	Abbreviation
(1)	Glutamic acid	
(2)		val
(3)		ser
(4)	Leucine	
(5)	Cysteine	
(6)	Histidine	
(7)		arg
(8)		pro
(9)	Methionine	
(10)	Phenylalanine	

7. For the amino acids in question 6, indicate whether they are acidic, basic, or neutral. If they are neutral, further classify the side chain as polar or nonpolar based on your knowledge of organic functional groups.

8. What type of functional group is formed when two amino acids are joined together. Are all amino acids able to form this functional group? What special name is given to the new bond formed?

9. Write the dipeptide structure formed when glycine is joined to alanine. Point out the peptide linkage. Name the dipeptide.

10. Name the following pentapeptides.
    (1) phe-thr-pro-leu-gly
    (2) his-gly-ala-tyr-val

11. An octapeptide was known to contain two proline residues and two leucine residues. The following peptides were obtained after partial hydrolysis. Determine the sequence.

    ala-pro-his, pro-leu, ser-leu, his-pro, pro-leu-val-ser

12. Secondary structure of proteins is held together by hydrogen bonding between the oxygen of the $\overset{\backslash}{\underset{/}{C}}=O$ and the hydrogen of the $H-\overset{/}{\underset{\backslash}{N}}$ groups in the polypeptide chain. Which amino acid is likely to disrupt secondary structure and why?

13. What is the only amino acid which forms a covalent link with its side chain?

14. Why is running a high fever for several days dangerous to a person's health (or even a very high fever for several hours?)

15. On what basis are proteins separated by electrophoresis?

16. There are three common electrophoresis procedures. Which of these procedures would be best to separate the following protein mixtures? (standard, SDS, or isoelectric focusing).
    (a) proteins of similar molar mass but different overall charges
    (b) proteins of different molar mass and different charges
    (c) proteins of similar charge but different molar mass

17. Match the term on the left with a phrase from the list on the right.

    (1) ninhydrin test
    (2) Sanger
    (3) peptide configuration
    (4) elements found in proteins
    (5) cysteine
    (6) Biuret test
    (7) chromatographic separation techniques
    (8) xanthroproteic reaction
    (9) denaturation

    a. changes in protein properties without hydrolysis
    b. C, H, O, S, N
    c. concentrated nitric acid
    d. α-helix and β-pleated sheet
    e. disulfide bonding
    f. violet color with copper sulfate
    g. amino acid sequence in insulin
    h. thin layer, paper, column
    i. blue color except with proline and hydroxyproline

18. Quarternary structure of hemoglobin is important in the (1) oxygen/hydrogen transport system of the blood. When the hemoglobin, composed of (2) _____ subunits, binds one molecule of oxygen to itself, its conformation changes to (3) discourage/facilitate the binding of (4) _____ additional oxygen molecules. As hemoglobin moves from the lungs to cells needing oxygen, one oxygen is removed. The protein (5) _____ changes again and the (6) _____ remaining oxygen molecules are then more easily removed. The presence of a (7) _____ structure allows the binding/removal of one oxygen molecule to control the binding/removal of three other oxygen molecules.

19. Glycine is considered a neutral amino acid and yet its isoelectric point is at a pH of 6.0. Explain.

RECAP SECTION

Chapter 30 has given us insights into one of the many fascinating areas of biochemical research. Proteins are composed from a relatively small number of building blocks, the amino acids, yet their structure and function are very complex. Scientists have managed to develop techniques used to separate proteins and, in some cases, synthesize or replicate proteins in the laboratory. Perhaps the best known function of proteins are their role as enzymes but proteins have a large variety of other purposes. Research has allowed scientists to begin to see how protein structure, enzyme reactivity, and genetics are interrelated. These molecules are very complex in their structure, yet there is an orderliness about their synthesis that is fascinating. Chapter 32 takes us further in the study of the genetic molecules themselves and how their information is passed on to new organisms.

ANSWERS TO QUESTIONS

1. fish, chicken, nuts, cheese, eggs, beans

2. The acidic, basic, or neutral classification refers to the side chain fo the amino acid. If an amino acid contains a carboxylic acid on the side chain, it is termed acidic. If the side chain contains an amine group, the amino acid is termed basic.

3. Yes, there are several neutral amino acids that have ionizable side chains. Cysteine and tryptophan are two examples. However, at physiological pH, these groups are generally in a neutral form.

4. (1) b   (2) e   (3) a   (4) f
   (5) d   (6) c   (7) g

5. (1) $$\text{CH}_3\text{-CH(CH}_3\text{)-CH(}^+\text{NH}_3\text{)-COOH}$$

   (2) $$\text{CH}_3\text{-CH(CH}_3\text{)-CH(NH}_2\text{)-COO}^-$$

6. (1) glu   (2) Valine   (3) Serine   (4) leu
   (5) cys   (6) his   (7) Arginine   (8) Proline
   (9) met   (10) phe

7. (1) acidic   (2) neutral, nonpolar   (3) neutral, polar
   (4) neutral, nonpolar   (5) neutral, polar   (6) basic
   (7) basic   (8) neutral, nonpolar   (9) neutral, nonpolar
   (10) neutral, nonpolar

8. An amide bond is formed between the carboxylic acid of one amino acid and the amine group of another amino acid. All amino acids are able to form this link because they all contain a primary or secondary amine group (tertiary amines cannot form amide bonds). When an amide bond is formed between two amino acids, the new bond is also called a peptide bond or peptide link.

9. $$\text{NH}_2\text{-CH}_2\text{-C(=O)-NH-CH(CH}_3\text{)-COOH}$$ glycyl alanine (peptide linkage)

10. (1) phenylalanylthreonylprolylleucylglycine
    (2) histidylglycylalanyltyrosylvaline

11. The octapeptide would be

    ala-pro-his-pro-leu-val-ser-leu

12. Proline disrupts secondary structure. Proline is the only amino acid which has a secondary amine and therefore when it forms an amide bind, the nitrogen does not have an attached hydrogen. The lack of an N—H in the polypeptide chain does not allow for hydrogen bonding which in turn tends to disrupt secondary structure.

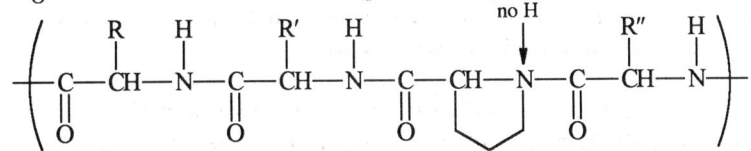

example of portion of polypeptide showing proline

13. Cysteine can form a disulfide bond between the side chains of two cysteine molecules.

14. Protein structure is closely linked to function. If the structure of the protein (1°, 2°, 3° or 4°) is disrupted, the protein ceases to function properly. Change in temperature may lead to protein denaturation. A high fever may cause the proteins to cease functioning properly because they are denatured. Because primary structure is not affected during denaturation, it is often reversible.

15. Electrophoresis separates proteins based on size and charge differences.

16. (a) isoelectric focusing (separates based on charge)
    (b) standard electrophonesis (separates by charge and size)
    (c) SDS (masks charge so separates based on size)

17. (1) i   (2) g   (3) d   (4) b   (5) e
    (6) f   (7) h   (8) c   (9) a

18. (1) oxygen   (2) four   (3) facilitate   (4) three
    (5) conformation or shape (6) three   (7) quarternary

19. Glycine's pH at its isoelectric point is acidic because the carboxyl group(—COOH) is more ionized than the amino group (—NH$_2$).

CHAPTER THIRTY-ONE

Enzymes

SELF-EVALUATION SECTION

1. Fill in the blank space or circle the appropriate response.

   The catalysts of biochemical reactions are called (1) _____. These molecules are generally (2) <u>proteins/metals</u>, but recently other compounds known as (3) _____ acids have been found to function as enzymes. Catalytic function depends on an enzyme's (4) _____ dimensional structure. Biological reactions must overcome an energy barrier called (5) _____ energy in order to occur in the cellular environment. Enzymes can be simple or conjugated with a protein part called (6) _____ and a nonprotein part (7) _____. Both parts together form the functioning enzyme or (8) _____.
   For some enzymes a metallic ion such as Ca^{2+} or Mg^{2+} is required. These ions are called (9) _____. The substance acted on by an enzyme is called a (10) _____.

2. There are six main classes of enzymes—lyases, ligases, isomerases, transferases, oxidoreductases, and hydrolases. Match the class of enzyme with the statement that characterizes its function.

   (1) catalyze the oxidation-reduction reaction between two substrates _____
   (2) catalyze the transfer of a functional group between two substrates _____
   (3) catalyze the interconversion of stereoisomers and structural isomers _____

(4) catalyze the removal of groups from substrates
by mechanisms other than hydrolysis _____

(5) catalyze the linking together of two compounds with
the breaking of a pyrophosphate bond in ATP _____

(6) catalyze the hydrolysis of esters, carbohydrates,
and proteins _____

3. Identify the following reactions as being catalyzed by which type of enzyme from the six main classes:

(1) Glucose $\longrightarrow$ fructose
$C_6H_{12}O_6 \qquad C_6H_{12}O_6$ _____

(2) Lactose $+ H_2O \longrightarrow$ Glucose $+$ Galactose
$C_{12}H_{22}O_{12} \qquad\qquad C_6H_{12}O_6 \quad C_6H_{12}O_6$ _____

(3) Amino acid $+$ tRNA $+$ ATP $\longrightarrow$
aminoacyl $-$ tRNA $+$ AMP $+ H_4P_2O_7$ _____

(4) $CO_2 + H_2O \longrightarrow HCO_3^- + H^+$ _____

4. Which of the following statements are correct?

(1) Enzyme specificity means that a particular enzyme only catalyzes a reaction for a specific substrate.
(2) Enzymes reduce the concentration of biological reactants at random sites by adsorption.
(3) Hydrolases catalyze the hydrolysis of esters, etc.
(4) The active site on an enzyme's surface is rigid and inflexible.
(5) Enzymes are under careful cellular control.
(6) The "lock and key" hypothesis has been proposed to help explain an enzyme's specificity.
(7) The most important commercial use of enzymes is as stain remover in detergents.
(8) During an enzyme-catalyzed reaction, the reactant must pass through a state termed the coordination state in order to be converted into a product.
(9) Two of the most common ways to increase an enzyme-catalyzed reaction rate are to increase the reactant concentration and decrease the temperature.
(10) Turnover number measures how many substrate molecules one enzyme molecule can react with in a given period.
(11) The strain hypothesis states that a reactant molecule is impelled to change shape to fit the binding site.
(12) Feedback inhibition affects enzymes at the end of the molecular assembly line.

5. How are enzymes able to increase the rate of a reaction?

6. The enzyme pyruvate carboxylase contains a covalently attached molecule named biotion whose structure is given below. What are the holoenzyme, coenzyme, and apoenzyme in this situation?

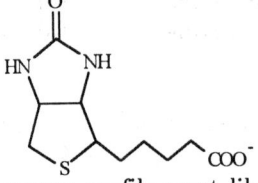

7. (a) Which energy profile most likely represents the enzyme-catalyzed version of the same overall reaction? Why?

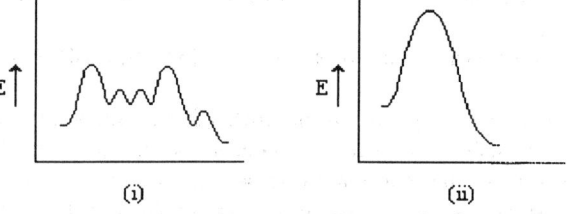

(i)   (ii)

(b) What are two other methods used to increase the rate of a reaction?

8. What does lactase catalyze? What type of reaction is it? (addition, . . . ?) What type of bond is broken during the reaction catalyzed by lactase?

9. What are the advantages of using enzymes in industry versus other chemicals or non-enzymatic catalysts? Give an example.

10. Why are enzymes not used more frequently to catalyze industrial chemical processes?

RECAP SECTION

Chapter 31 describes the structure and function of chemical compounds that are vital for all of life's processes. Scientists have gained a great deal of insight into how enzymes work and what can influence their efficiency. Many of the exciting scientific discoveries of the last 90 years have involved work with enzymes. Chemists today are exploring various commercial applications for enzyme technology, and the field remains a rewarding area for creative work.

ANSWERS TO QUESTIONS

1. (1) enzymes (2) proteins (3) ribonucleic (4) three
   (5) activation (6) apoenzyme (7) co-enzyme (8) holoenzyme
   (9) activators (10) substrate

2. (1) oxidoreductases (2) transferases (3) isomerases (4) lyases
   (5) ligases (6) hydrolases

3. (1) isomerase (2) hydrolase (3) ligase (4) transferase

4. The correct statements are: (1), (3), (5), (6), (10), (11)

5. Enzymes generally do one or both of the following: they bring the reactants close together thus increasing the effective concentration and therefore speeding up the reaction or the enzyme positions the reactants so they are set up to break and make the necessary bonds. The former is called proximity catalysis and the latter process is called the productive binding hypothesis. An enzyme may also speed up a reaction by forcing a reactant to be strained in order to fit in the binding site and thus is more reactive (strain hypothesis).

6. The protein portion of the enzyme, the apoenzyme, is pyruvate carboxylase. The coenzyme is biotin and together the pyruvate carboxylase and biotin comprise the holoenzyme, the entire functioning enzyme.

7. (a) Energy profile (i) represents the enzyme-catalyzed reaction. The lower the activation energy, the faster the reaction. Also, enzyme catalysis generaly alters the path of the reaction although not the initial or final states.
   (b) Two other methods of increasing the rate of the reaction are increasing the concentration of the reactants or increasing the temperature of the reaction.

8. Lactase catalyzes the hydrolysis of lactose. A glycosidic bond (an acetal bond) is broken during the hydrolysis.

9. Enzymes are very specific and often work under less harsh conditions than processes which do not use enzymes. Enzyme-catalyzed reactions often produce little or no toxic by-products and, because they usually have a high turnover, generally require smaller quantities of catalyst. Enzymes used as detergent additives can degrade protein stains and operate at lower wash temperatures.

10. Enzymes specific for the process may not be known or available. Enzymes are difficult to synthesize and are very specific. They are sometimes difficult to isolate from other proteins. (The production problems may sometimes be solved by genetic engineering.)

CHAPTER THIRTY-TWO

Nucleic Acids and Heredity

SELF-EVALUATION SECTION

1. Fill in the blank space or circle the correct response.

    The great variety of proteins found in any organism does not arise by chance. The substances that direct the course of protein synthesis are known as (1) _____. The two major types are known by abbreviations, (2) _____ and (3) _____. There are several major differences between these types of nucleic acids. RNA contains the sugar (4) _____, while DNA contains (5) _____. DNA is a double-stranded (6) _____, while RNA is only (7) _____ stranded. In addition, one of the pyrimidine bases is different. DNA contains the base (8) _____ and RNA contains (9) _____. The term nucleoside refers to either the purine or pyrimidine base joined with a (10) _____ molecule. There are (11) 10/15 such nucleotides, each with a specific abbreviation. Thus, A stands for (12) _____, dG stands for (13) _____, T stands for (14) _____, and dU stands for (15) _____. In nucleic acids, the actual building blocks are phosphate esters of nucleosides. These molecules are called (16) _____. Again, there are specific abbreviations. AMP stands for (17) _____, ADP for (18) _____, and ATP for (19) _____. These last two nucleotides are extremely important biological compounds as easily obtainable chemical energy residues in their hydolyzable (20) _____ bonds. Thus, when ATP is hydrolyzed to ADP and inorganic phosphate via an enzyme catalyzed reaction, a considerable amount of energy is released.

In the double-stranded helix of DNA, it has been found that (21) <u>complementary/dissimilar</u> pairing occurs between purine and pyrimidine bases. dA pairs with (22) _____ and dG pairs with (23) _____. The genetic information contained in the cell's genes is found to be the exact sequence of nucleotides along the DNA strands. When the DNA replicates itself during normal cell division called (24) _____, the two daughter cells have the same double-stranded DNA sequences as the original cell. During sex cell production, called (25) _____, only half the genes or DNA sequences are present in the new cells. When fertilization takes place the zygote has a normal complement of genes, half from the mother and half from the father. Each gene, which directs the synthesis of one protein, contains what is termed the genetic code. Each codon, for a specific amino acid, is composed of (26) _____ nucleotides in sequence. The intermediate compound, which carries the information from the nucleus (site of the DNA) to the ribosomes (site of protein synthesis), is a type of ribonucleic acid called (27) _____. The ribonucleic acid, which brings various amino acids to the ribosomes, is called (28) _____ and has the shape of a (29) _____. Each loop of a particular *t*-RNA has a purpose.

If, for some reason, the DNA structure is changed or a breakdown occurs in the protein synthesis process, a (30) _____ may occur. Most severe cases result in an organism's death, but sometimes the organism survives. For example, sickle cell anemia results from (31) <u>two/one</u> change(s) in an amino acid in one chain of hemoglobin.

2. Match the term on the left with a phrase or term on the right. Some choices may be used more than once.

   (1) guanine
   (2) cytosine
   (3) transfer RNA
   (4) uracil
   (5) messenger RNA
   (6) ribosomal RNA
   (7) adenine
   (8) thymine
   (9) genome
   (10) oncogenes
   (11) AUG and GUG
   (12) GTP

   a. purine base
   b. causes cells to become cancerous
   c. pyrimidine base
   d. carries genetic information from DNA to ribosomes
   e. codons for the start of protein synthesis
   f. brings amino acids to the ribosomes for incorporation into protein
   g. nuclotide that is primary source of energy for protein synthesis
   h. part of structure at which protein synthesis occurs
   i. sum of all hereditary material contained in a cell

3. What is the difference between a nucleioside and a nucleotide?

4. What are three major structural differences betwen DNA and RNA?

5. What is the major difference in function between DNA and RNA?

6. Why are the nucleic bases called bases?

7. Which three scientists made the most significant contributions to deducing the structure of DNA? For which two is the double-stranded model named?

8. What is the difference between mitosis and meiosis?

9. What is the complementary strand to CGATTACCGAT? Is this a strand of DNA or RNA? How do you know?

10. A segment of DNA has the structure TAGTACGGTAACAAAAGGTCGCCGATC. What is the sequence of the polypeptide chain an mRNA reading this strand will synthesize?

11. Try to fill in the boxes or spaces with the correct terms without using the text.

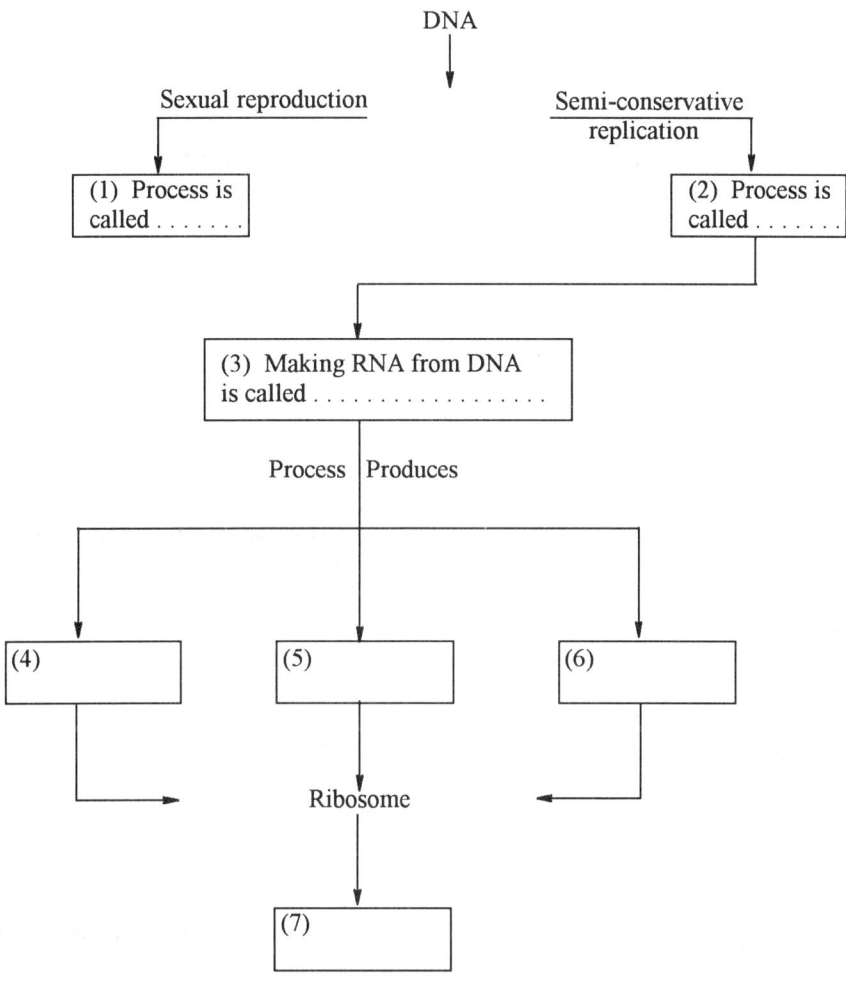

The ribosomes are the site of (7) _____ synthesis, a process called (8) _____. A primary function of the tRNA is to bring (9) _____ to the ribosomes for incorporation into proteins. The messenger RNA contains the information or (10) _____ in the form of three (11) _____. Some amino acids have only one codon each but for other amino acids the code is (12) _____.

RECAP SECTION

Chapter 32 is an excellent introduction to a very complex and fascinating subject. The science of genetics and the biochemical basis for heredity and the genetic code are current areas of research today. Much has been learned in the last 25 years about the process of protein synthesis and several genetic diseases have been explained and are now treatable. The future promises other problem areas now that researchers are able to alter the genetic material and induce various types of mutations. Cloning of organisms is another concern. An informed public is of great importance in the area of DNA research.

ANSWERS TO QUESTIONS

1. (1) nucleic acids
    (2) DNA
    (3) RNA
    (4) ribose
    (5) 2-deoxy-D-ribose
    (6) helix
    (7) single
    (8) thymine
    (9) uracil
    (10) sugar
    (11) 10
    (12) adenosine
    (13) deoxyguanosine
    (14) thymidine
    (15) deoxyuridine
    (16) nucleotides
    (17) adenosine monophosphate
    (18) adenosine diphosphate
    (19) adenosine triphosphate
    (20) phosphate
    (21) complementary
    (22) dT
    (23) dC
    (24) mitosis
    (25) meiosis
    (26) three
    (27) messenger RNA
    (28) transfer RNA
    (29) cloverleaf
    (30) mutation
    (31) one

2. (1) a   (2) c   (3) f   (4) c   (5) d   (6) h
    (7) a   (8) c   (9) i   (10) b (11) e (12) g

3. A nucleoside is composed of the sugar and a nucleic base. A nucleotide has a phosphate ester in addition to the sugar and the base.

4. RNA and DNA differ structurally primarily in four ways. DNA uses deoxyribose sugar, while RNA uses ribose sugar. DNA generally exists as a double strand (double helix) while RNA is mainly a single polynucleotide strand. DNA contains the bases (A, C, G and) T while RNA contains (A, C, G and) U. Lastly, RNA sometimes may have significant amounts of unusual bases.

5. DNA stores the genetic code while RNA allows the transfer and expression of the genetic characteristics.

6. The pyrimidine and purine structures are nitrogen heterocycles and will react with H^+ to make a solution more basic.

7. Maurice Wilkins, James Watson, and Francis Crick shared the 1962 Nobel prize in medicine and physiology for their studies of DNA. The double-stranded helix is known as the Watson-Crick model of DNA.

8. Both processes are related to cell reproduction. In mitosis, daughter cells have the same number of chromosomes as the parent cell. In meiosis, daughter cells have half the number of chromosomes as the parent (meiosis generally limited to reproductive cells).

9. The complementary strand is GCTAATGGCTA. The strand is from DNA because the base T was involved. RNA uses uracil not thymine.

10. The mRNA strand will be complementary to the DNA strand. So the sequence of bases on mRNA will be

    AUC AUG CCA UUG UUU UCC AGC GGC UAG.

    The bases were written in sets of three to easily identify the genetic code for the various proteins. Using Table 32.3 in the text, the polypeptide chain should be

    Ile–Met–Pro–Leu–Phe–Ser–Ser–Gly.

    The last codon indicated termination of the chain.

11. (1) meiosis (2) mitosis (3) transcription
    (4) mRNA (5) rRNA (6) tRNA
    (4), (5), (6) in any order
    (7) protein (8) translation (9) amino acids
    (10) code (11) nucleotides (12) redundant

CHAPTER THIRTY-THREE

Nutrition

SELF-EVALUATION SECTION

1. Fill in the missing places in the following table.

Digestive Fluid Source	Fluid Name	Principal Enzymes
Salivary glands	(1) _____	(2) _____
(3) _____	(4) _____	Pepsin, rennin, gastric lipase
(5) _____	Pancreatic juice	Trypsin, chymotrypsin
(6) _____	(7) _____	Bile salts
Glands in duodenum	(8) _____	Finishing enzymes

2. Match the digestive fluid with the type of food digestion catalyzed or facilitated by the particular fluid.

   (1) gastric
   (2) bile juice
   (3) pancreatic juice
   (4) saliva

   a. digestion of all three kinds of food
   b. starch hydrolysis, lubrication
   c. protein digestion
   d. fat digestion

3. List the six nutrient groups.

4. List the five food groups.

5. Name at least 3 of the major elements and 3 of the trace elements.

6. Match the phrases or descriptions on the right with the terms on the left.

   (1) water
   (2) vitamins
   (3) marasmus
   (4) RDA
   (5) obesity
   (6) Kwashiorkor
   (7) cellulose
   (8) complete protein
   (9) diet

   a. recommended dietary allowance
   b. supplies all essential amino acids
   c. excess calories
   d. calorie deficiency
   e. food and drink we consume
   f. dietary fiber
   g. components of enzyme systems
   h. body fluids, reaction solution
   i. protein deficiency

7. To lose one pound of body fat requires an energy expenditure of 3500 kcal greater than the energy intake.

   If a person could realistically reduce their energy allowance by 500 kcal per day, how many days would be required to lose 30 pounds?

8. List four purposes of food additives and give an example of an additive for each purpose.

9. Which nutrient provides the most energy per unit mass?

10. (1) What is the structural difference between a saturated fat and an unsaturated fat?

    (2) What type of reaction converts an unsaturated fat to a saturated fat?

11. What two changes should a person make related to his food intake if he desires to lose weight in a healthy manner?

12. Use the following information to answer the questions below.

	Tuna (in water)	Peanut Butter	Cheese	Strawberry Jam
Serving size	2 oz ($\frac{1}{4}$ cup)	2 Tbsp	1 oz	1 Tbsp
Calories	60	190	110	50
Fat Calories	5	140	80	*
Amount/Serving (%DV)[+]				
Total Fat	0.5g (1%)	17g (26%)	9g (14%)	0g (0%)
Sat. Fat	0g (0%)	3g (17%)	5g (25%)	*
Cholesterol	30mg (10%)	0mg (0%)	30mg (10%)	*
Sodium	250mg (10%)	140mg (6%)	180mg (8%)	10mg (0%)
Total Carbohydrates	0g (0%)	5g (2%)	1g (0%)	13g (4%)
Fiber	0g (0%)	2g (8%)	0g (0%)	*
Sugars	0g	2g	0g	9g
Protein	13g (23%)	9g (16%)	7g (12%)	0g
Vitamin A	(0%)	(0%)	(6%)	*
Vitamin C	(0%)	(0%)	(0%)	*
Calcium	(0%)	(0%)	(20%)	*
Iron	(2%)	(2%)	(0%)	*

+   (%DV) based on a 2000 cal diet.
*   not a significant source

Jack, Mary, and Jane made a tuna sandwich, a peanut butter and jam sandwich, and a cheese sandwich respectively. Jack used $\frac{1}{2}$ cup of tuna. Mary used $\frac{1}{2}$ Tbsp peanut butter and $2\frac{1}{2}$ Tbsp jam. Jane made her sandwich with 4 oz of cheese. Assuming all three sandwiches contained the same amount of bread and no other condiments:

(1) Which person got the most energy from their meal?

(2) Which lunch provided the most macronutrients by mass?

(3) Which lunch(es) would a nutritionist say is too high in fat?

(4) What is lacking in all three sandwiches?

(5) Which lunch has the largest amount of empty calories?

(6) Which sandwich filler provides the most protein energy and how much is provided?

(7) Which meal is best for a person suffering from occasional night blindness?

(8) Which meal has the highest percentage of the major elements?

RECAP SECTION

Chapter 33 discusses the biochemical aspects of proper nutrition and some of the consequences of improper diet. The relationships between various nutrients in a balanced diet are described as is the process of digestion. These topics are of general interest to us all and of special importance to persons in the health fields.

ANSWERS TO QUESTIONS

1. (1) saliva (2) amylase (3) stomach glands (4) gastric juice
   (5) pancreas (6) liver (7) bile (8) intestinal juice

2. (1) c (2) d (3) a (4) b

3. carbohydrates, lipids, proteins, water, minerals, vitamins
4. milk products, vegetables, fruits, cereal products, meats

5. Na, K, Ca, Mg, Cl, P are the major elements.
   F, Si, Cr, Fe, Co, Ni, Cu, Zn are examples of the trace elements.

6. (1) h (2) g (3) d (4) a (5) c (6) i (7) f (8) b (9) e

7. 210 days or 30 weeks

   $$\frac{(30 \text{ pounds})(3500 \text{ kcal/pound})}{500 \text{ kcal/day}} = 210 \text{ days}$$

8. (1) enhance nutritional value—vitamins, minerals
   (2) preservative—sodium benzoate, calcium lactate, sorbic acid
   (3) anti-oxidant—BHA, BHT
   (4) improve appearance and flavor—pectin, propylene glycol

9. Fat provides the most energy per unit mass (9 kcal/g).

10. (1) An unsaturated fat contains alkene functional groups ($C=C$). Saturated fats contain no carbon-carbon double bonds.

    (2) Alkenes can be saturated by adding hydrogen in a process called hydrogenation. Hydrogenation is an addition reaction.

11. A person should reduce caloric intake moderately which is often achieved by reducing fat intake. He should, however, ensure that the foods chosen contain proper amounts of nutrients.

12. (1) Jane. The cheese sandwich provides 440 calories from the cheese. The fillings for the other sandwiches contained 120 calories and 362.5 calories. The more calories, the more energy is provided.
    (2) Jane. Macronutrients include carbohydrates, proteins, and fats. The cheese provided 68g macronutrients, the peanut butter and jam provided 58.25g macronutrients, while the tuna provided 27g of macronutrients. In terms of %DV, cheese again provided the highest total percentage.
    (3) The cheese sandwich provides 56% of the daily value of fat. However as a percentage of the meal, 73% of the calories came from fat which is considered too high (target 25–30%). The peanut better and jam also provided more that 30% of its calories from fat.
    (4) All three sandwiches are low in micronutrients. The cheese is the best source of micronutrients.

(5) Sugar provides empty calories. Mary's peanut butter and jam sandwich contains the most sugars.
(6) Cheese provided 28g protein which is about 118 calories.
(7) Persons who have a vitamin A deficiency often suffer from night blindness. The cheese sandwich provides the most vitamin A.
(8) The cheese sandwich has the highest percentage of the major elements (the only ones listed in the nutrition information are sodium and calcium).

CHAPTER THIRTY-FOUR

Bioenergetics

SELF-EVALUATION SECTION

1. What is meant by the term bioenergetics?

2. Which molecule is commonly thought of as being the energy source for the cell. Where is this energy stored?

3. Fill in the blank space or circle the correct response.

   The sum of all chemical changes that occur in a living organism is given the name (1) _____. Further, these changes have two aspects. When a process involves simple substances being synthesized into complex substances, a process is termed (2) _____. The opposite process, involving the breaking down of complex materials into simpler substances, is termed (3) _____. The raw biochemical materials used as sources for the metabolic energy processes are classed as (4) _____, (5) _____ and (6) _____. These energy sources originally were derived from the energy of (7) foodstuffs/sunlight by a process in green plants called (8) _____. Energy is made available to the cell by means of (9) displacement/oxidation-reduction reactions which involve the transfer of (10) _____. The organelle which is the site of most oxidation-reduction reactions is the (11) _____. The transport of electrons from one reaction site to another involves various coenzymes derived from B-

complex vitamins. The three most common coenzymes in electron transport are (12) _____, (13) _____ and (14) _____.
The released energy is stored in a phosphate anhydride bond in the molecule (15) _____.

4. Name two advantages to a cell for using ATP as a stored form of energy.

5. Identify the following phrases as applying to either substrate-level phosphorylation (SLP) or to oxidative phosphorylation (OP).

   (1) oxidation of reduced carbon to form high energy phosphate bonds _____.
   (2) direct use of oxidation-reduction reactions to form ATP _____.
   (3) site is mitochondria and involves electron transport system _____.
   (4) involves phosphorylated substrates _____.

6. List 4 different subcellular organelles.

7. Match the phrase or key words from the right-hand list with the terms on the left.

   (1) procaryote cell
   (2) eucaryote cell
   (3) liver
   (4) membrane-bound ribosomes
   (5) kidneys
   (6) mitochondria
   (7) chloroplasts
   (8) red blood cell
   (9) oxidative phosphorylation
   (10) photosynthesis

   a. component responsible for rapid oxygen transport
   b. site of photosynthesis
   c. production site of most cellular energy molecules
   d. produce proteins for secretion from the cell
   e. lacks distinct membrane-bound nucleus
   f. converts lactic acid to glucose
   g. membrane-bound organelles and nucleus
   h. transfer water from bloodstream to urine
   i. requires oxygen
   j. source of nearly all biochemical system energy

8. Fill in the blank space with an appropriate term.

   A good example of a coordinated cell function is the removal of carbon dioxide from muscle cells to the (1) _____ where it is exhaled. This process is tied to the release of oxygen from the transport molecule (2) _____ at a site of use, the muscle cells. During exertion, the muscle cells produce CO_2 and require O_2.

   Carbon dioxide reacts with amino groups on the (3) _____ molecule and releases hydrogen ions. The hydrogen ions promote the release of more oxygen from oxyhemoglobin to be used by the muscle cells. A second removal process for carbon dioxide also involves the production of hydrogen ions in the muscle cells. In the lungs, the hemoglobin is reoxygenated. Carbon dioxide is released by the reactions in the lungs and is exhaled.

The liver and the kidney also play important roles in maintaining the proper acid (hydrogen ion) level in the body. The liver replaces reduced carbon energy sources in the blood stream as muscle cells increase their activity. During hard work, muscle cells will function in an oxygen deficient environment temporarily and produce (4) _____ instead of carbon dioxide. The liver converts this product back to (5) _____ for recycling to the muscle cells. The kidneys, a prime regulator of water and ion balance, selectively transfer water and ions from the blood to the (6) _____ which is a slow process so that kidney function is important for the long term maintenance of correct body fluid pH values.

9. Match phrase in left-hand column with correct phrase in right-hand column:

<u>Energy use at muscle</u>   <u>How gas balance restored in the lungs</u>

(1) O_2 used   a. CO_2 exhaled
(2) CO_2 formed   b. more O_2 bound to Hb
(3) acidity increases   c. O_2 inhaled
(4) more O_2 released from HbO_2 d. acidity decreases

10. When the muscles use energy does the pH of the blood increase or decrease? Why?

11. Why is carbon monoxide lethal if breathed in continuously for a period of time?

12. Compare the oxygen binding curve for hemoglobin to that for myoglobin.
    (1) What percentage of oxygen is released when the pressure is reduced from 50 to 10 torr?
    (2) Which protein is better at storing oxygen?
    (3) Would you expect humans or whales to have a larger percentage of the protein you chose in (2)?

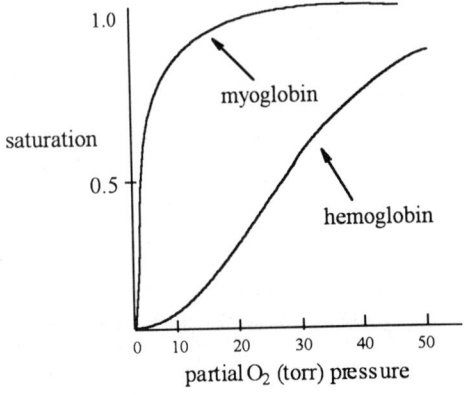

RECAP SECTION

Chapter 34 focuses on the many processes that are related to energy transformations in living organisms. Several of these topics have already been discussed briefly previously in the text. The importance of oxidation-reduction processes becomes obvious and the primary locations of many of the reactions within the cell is revealed. The function of the liver, the kidneys, the lungs and the muscles in maintaining the correct hydrogen ion, CO_2, and O_2 balance is a very important topic. This chapter reemphasizes that the application of basic chemical principles to biological problems is a worthwhile and exciting endeavor.

ANSWERS TO QUESTIONS

1. Bioenergetics refers to the processes by which energy is obtained, converted, distributed, and ultimately used by living organisms.

2. ATP. The energy is stored in the phosphate anhydride bonds.

3. (1) metabolism (2) anabolism (3) catabolism (4) carbohydrates
   (5) lipids (6) proteins (7) sunlight (8) photosynthesis
   (9) oxidation-reduction (10) electrons (11) mitochondria
   (12) NAD^+ (13) $NADP^+$ (14) FAD (15) ATP

4. (1) Energy stored in ATP is readily accessible through a simple hydrolysis reaction.
   (2) Energy from various types of reactions is funneled into one storage form, ATP. ATP is a common energy currency in a cell.

5. (1) SLP (2) OP (3) OP (4) SLP

6. Possibilities are: nucleus, ribosome, mitochondrion, chloroplasts, lysome, peroxisome, Golgi apparatus

7. (1) e (2) g (3) f (4) d (5) h (6) c (7) b (8) a (9) i (10) j

8. (1) lungs (2) hemoglobin (3) hemoglobin (4) lactic acid (5) glucose (6) urine

9. (1) c (2) a (3) d (4) b

10. The pH of the blood decreases (blood gets more acidic) because the carbon dioxide formed reacts with both hemoglobin and water to produce hydrogen ions.

11. Carbon monoxide has a greater affinity for hemoglobin than does oxygen and is not easily released by hemoglobin. The hemoglobin is unavailable to transport oxygen and a person dies.

12. (1) Hemoglobin releases close to 90% of the oxygen it bound at 50 torr with a 40 torr drop in partial pressure while myoglobin releases almost no oxygen.

(2) Myoglobin is better at storing oxygen because it needs a larger drop in pressure to release most of its oxygen.
(3) Whales have a larger amount of myoglobin because they need to hold their breath for longer periods of time and therefore need to store oxygen more efficiently than humans.

CHAPTER THIRTY-FIVE

Carbohydrate Metabolism

SELF-EVALUATION SECTION

1. What is the difference between anabolism and catabolism?

2. What is the difference between hypoglycemia and hyperglycemia?

3. What is the difference between aerobic and anaerobic processes?

4. Fill in the blank space or circle the appropriate response.

   In the body, glucose can be oxidized to carbon dioxide and (1) _____ with the trapping of the energy in the form of phosphate bond energy in the transfer molecule (2) _____. Or glucose can be stored as (3) _____ to function as a reserve.

   The anaerobic conversion of glucose to pyruvate is called the (4) _____ pathway. The process is called anaerobic because no (5) _____ is involved. Once pyruvate is formed it enters an aerobic cycle known as the (6) _____ cycle, which eventually extracts most of the chemical bond energy originally present in the glucose.

   Whatever the fate of glucose and other carbohydrate molecules during metabolism, each of the reactions is catalyzed by specific enzymes, and in many cases controlled by regulatory molecules called (7) _____ . These molecules are secreted by the ductless or (8) _____ glands and are synthesized by the body. The key molecule in the metabolic relationship between amino acids, fatty acids, and the citric acid cycle is (9) _____ .

5. Figure 35.3 shows the citric acid cycle. The citric acid cycle is termed an aerobic sequence but nowhere in the diagram does oxygen actually appear. Explain.

6. The diagram below represents an overview of carbohydrate metabolism. From the list of terms given, identify A, B, C, and D.

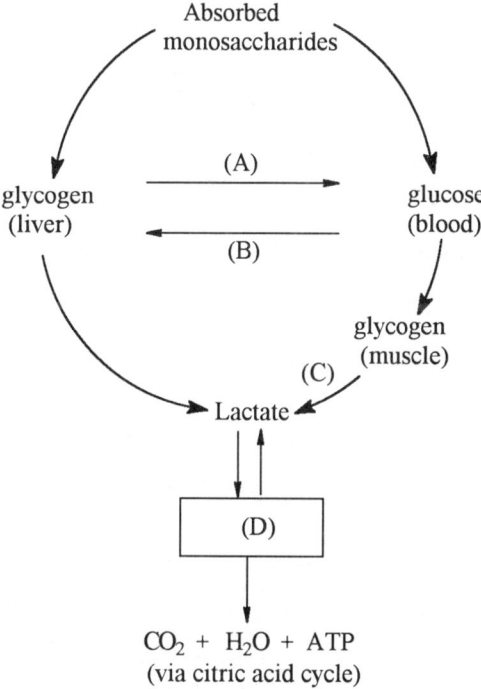

Terms to be used for diagram above and the diagram in question number 7 which follows:

(1) ketoglutaric acid
(2) glycolysis
(3) ATP
(4) acetyl-CoA
(5) glycogenesis
(6) glycine
(7) pyruvate
(8) amino acids
(9) glycogenolysis
(10) fatty acids
(11) NADH
(12) chlorophyll
(13) glucose
(14) CO_2

7. The diagram below represents a simplified illustration of the interrelationship of carbohydrates, fats, and proteins in the body.

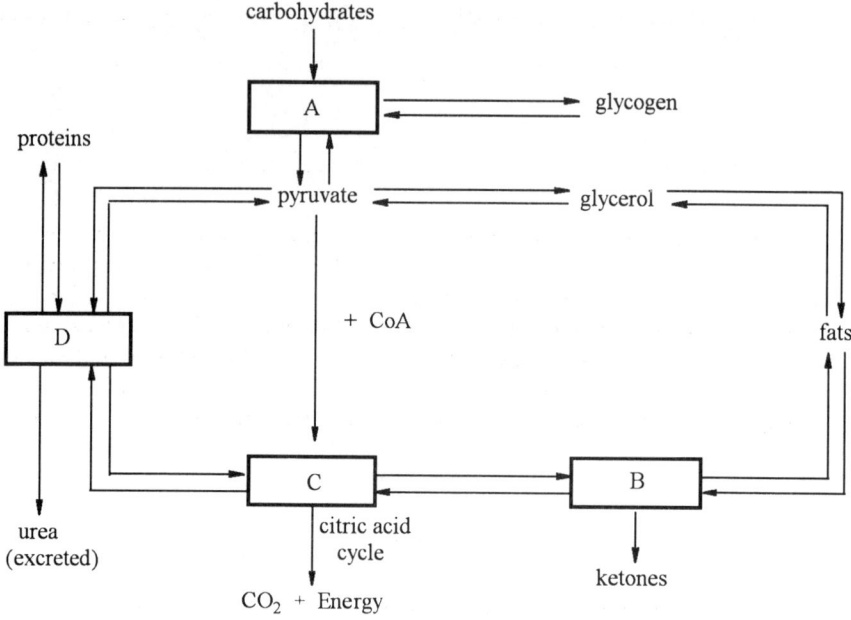

8. Match the terms on the left with one or more items from the list on the right.

   (1) Hyperglycemia
   (2) Hypoglycemia
   (3) Renal threshold
   (4) Insulin
   (5) Epinephrine

   a. reduces blood glucose
   b. below fasting level
   c. increases blood glucose
   d. glucose above fasting level
   e. increases rate of glycogen breakdown
   f. increases rate of glycogen formation
   g. glucose appears in urine
   h. 90-140 mg glucose per 100 mL of blood

9. (1) What is the structure of glucose?

   (2) Oxidation of which carbon would yield the least amount of energy and why?

10. Using the following schematic overview of the citric acid cycle, and without referring to the text, at which points along the cycle is CO_2 lost?

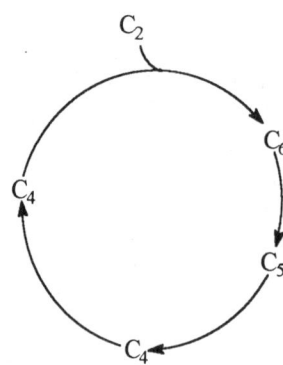

11. (a) Do you think a diabetic would become hypoglycemic or hyperglycemic if he took insulin and then forgot to eat?

    (b) What should a diabetic do if they are experiencing the other condition?

RECAP SECTION

Chapter 35 continues where the previous chapter left off. In Chapter 34, the digestive processes were summarized. In this chapter, the products of digestion, the monosaccharides are utilized to produce energy or stored for later use. The regulation of carbohydrate metabolism by hormones and vitamins are of particular interest. Molecules such as glucose and glycogen are of special concern because of their importance as metabolic intermediates and energy reserves. Carbohydrate metabolism plays a significant role in normal growth and development. A knowledge of metabolism is an integral part of many biological careers.

ANSWERS TO QUESTIONS

1. Anabolism refers to biosynthetic processes while catabolism refers to degradative processes.

2. Both terms refer to the level of glucose in the blood. Hypoglycemia is when the blood glucose level is low (below normal fasting levels). When the blood glucose level is high (above normal), hyperglycemia exists.

3. Aerobic processes require free oxygen or respiratory oxygen while anaerobic processes do not require oxygen.

4. (1) water
   (2) ATP
   (3) glycogen
   (4) Embden-Meyerhof
   (5) oxygen
   (6) citric acid
   (7) hormones
   (8) endocrine
   (9) acetyl-CoA

5. The citric acid cycle depends on electron transport chain which utilizes oxygen and therefore the citric acid cycle is termed aerobic?

6. A-9, B-5, C-2, D-7

7. A-13, B-10, C-4, D-8

8. (1) d, h   (2) b   (3) g   (4) a, f   (5) c, e

9. (1)
```
       O=C—H
   H ──┼── OH
  HO ──┼── H
   H ──┼── OH
   H ──┼── OH
       CH₂OH
```

(2) The oxidation of the aldehyde carbon should yield the least amount of energy because it is already the most highly oxidized carbon in glucose.

10. CO_2 is lost when the six-carbon molecule is converted to the five-carbon molecule and another CO_2 is lost when C_5 goes to C_4.

11. (a) If a diabetic took insulin and then forgot to eat, he would become hypoglycemic. The insulin would reduce the blood-glucose levels which unless replaced by food, would cause a person to be hypoglycemic.

(b) If a diabetic is experiencing hyperglycemia, his blood-glucose levels are too high and he should take insulin.

CHAPTER THIRTY-SIX

Metabolism of Lipids and Proteins

SELF-EVALUATION SECTION

1. Fill in the blank space or circle the appropriate response.

   The biological oxidation of fatty acids has been investigated since the early 1900's. Based on experimentation, Franz Knoop suggested that the chain was shortened by (1) _____ atoms at a time. This suggestion has been verified by other scientists. The carbon chain fragments can then feed into the Kreb's cycle and yield high-energy phosphate in the form of (2) _____ .

   The digestion of proteins provides alpha amino acids, which are absorbed through the intestinal wall into the blood stream. Amino acids undergo a variety of reactions while maintaining an equilibrium with the amino acid pool. When the amount of nitrogen excreted equals the nitrogen intake, a person is said to be in (3) _____ . A growing person should take in more than is excreted, and this is termed a (4) _____ nitrogen balance. On the other hand, an individual who is fasting or undergoing starvation will be in a (5) _____ nitrogen balance. Absorbed amino acids are usually incorporated into body protein. They also can lose the amino group to become an alpha keto acid or undergo transamination. Transamination is an important reaction linking amino acids to the metabolic pathways of carbohydrates and lipids.

2. List the possible metabolic fates of amino acids in humans.

3. Complete the following reactions of amino acids. Name the amino acid reactant in both reactions.

   (1) C$_6$H$_5$—CH$_2$—CH(NH$_2$)—COOH + H$_2$O + O$_2$ $\longrightarrow$

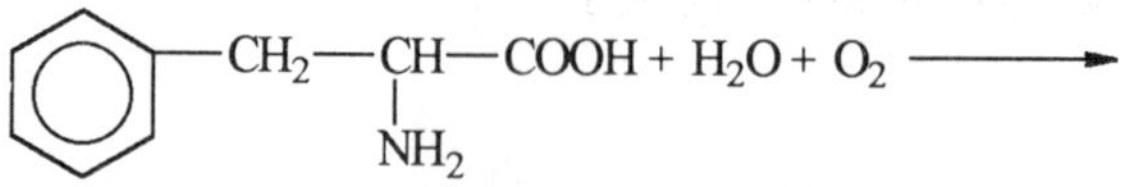

   oxidative deamination

   (2) CH$_2$(NH$_2$)—COOH + HOOC—CH$_2$CH$_2$C(=O)—COOH $\xrightarrow{\text{transaminase}}$

4. Identify the terms or phrases given as referring to lipogenesis (L) or fatty acid oxidation (O).

   (1) catabolic pathway
   (2) cytoplasm
   (3) malonyl CoA plays a role
   (4) anabolic pathway
   (5) mitochondria
   (6) fatty acid chain linked to ACP
   (7) fatty acid chain linked to CoA

5. How many acetyl CoA molecules can be produced by oxidation of the following fatty acids?

   (1) linolenic
   (2) stearic
   (3) myristic
   (4) capric
   (5) arachidic
   (6) caproic

6. Match the name of the biochemical process with the simplified reaction equation given:

   (1) D-glucose $\longrightarrow$ 2 pyruvic acid

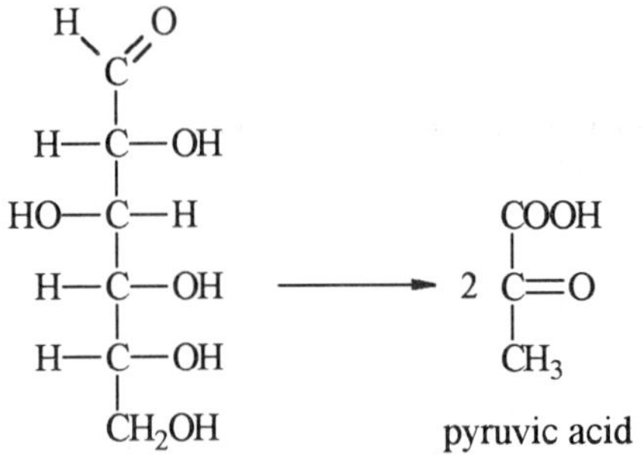

   (2) phosphoenolpyruvate and ADP $\longrightarrow$ ATP + pyruvate
   (3) Acetyl-CoA $\xrightarrow{[O]}$ CO$_2$ + H$_2$O + Energy
   (4) ADP + Pi + NADH + FADH$_2$ $\xrightarrow{[O]}$ H$_2$O + ATP
   (5) 6 CO$_2$ + 6 H$_2$O $\xrightarrow{\text{light}}$ C$_6$H$_{12}$O$_6$ + 6 O$_2$

   Process Names

   a. substrate level phosphorylation
   b. electron transport-oxidative phosphorylation
   c. photosynthesis
   d. Embden-Meyerhof pathway
   e. citric acid cycle

7. (1) The hydrolysis of a piece of the side chain of which amino acid yields urea?
   (2) What functional group is left on the resulting amino acid?

8. Lipogenesis and fatty acid oxidation are not the reverse of each other. However, there are some features which may lead some to believe the processes are the opposite of each other. What are these features?

9. What is the difference between a glucogenic amino acid and a ketogenic amino acid?

10. Arrange the following steps in the order they occur in beta oxidation:

    hydration, oxidation, cleavage, activation, oxidation

11. How is the majority of nitrogen transferred from one cell to another via the bloodstream?

12. (1) Which of the following is (are) most likely to be naturally occurring?
    (i) $CH_3CH_2CH_2CH_2CH_2CH_2CH_2CH_2CH_2CH_2CH_2CH_2CH_2CH_2COOH$
    (ii) $CH_3CH_2CH_2CH_2CH_2CH_2CH_2CH_2CH_2CH_2CH_2CH_2CH_2CH_2CH_2CH_2COOH$
    (iii) $CH_3CH_2CH_2CH_2CH_2CH_2CH_2CH_2CH_2CH_2CH_2CH_2CH_2CH_2CH_2CH_2CH_2COOH$
    (2) Why?
    (3) How many inorganic phosphate molecules would be released as a direct result of beta oxidation of the molecule(s) you chose in (1)?

13. Which amino acid is formed as a result of the following transamination? What other product is formed?

    $$CH_3CHCH_3 \atop \underset{}{CH_2CCOOH} \atop \underset{}{\overset{\|}{O}} \quad + \quad NH_2CH_2CH_2CH_2CH_2CH\underset{NH_2}{-}COOH \longrightarrow$$

14. Which of the following items draw amino acids away from the amino acid pool? Which items contribute amino acids to the pool? Are any of the items two way processes?

    amino acid pool

    (1) amino acids from transamination
    (2) food
    (3) excreted nitrogen
    (4) non-protein nitrogen compounds
    (5) tissue protein
    (6) positive nitrogen balance

15. Fill in the blank space or circle the appropriate resonse.

ATP (1) is/is not produced directly during β-oxidation of fatty acids but is formed when the reduced coenzymes (2) _____ and (3) _____ are oxidized by the mitochondrial electron system in concert with (4) _____ _____ . Fatty acid oxidation is (5) aerobic/anaerobic because FAH_2 and NADH can only be oxidized when (6) _____ is present.

RECAP SECTION

Chapter 36 is the final chapter in the discussion of biochemistry. This chapter concludes the exploration of the metabolic processes begun in the preceding chapters. Redox reactions were important in nutrition, digestive processes and continue to be so in fatty acid and amino acid metabolism. The involvement of acetyl-CoA and the regulatory molecules in fatty acid and amino acid metabolism has been of great interest in the scientific community. It is appropriate that the last few chapters deal with the topic of metabolism of the three major sources of chemical energy. All nonphotosynthetic organisms, including humans, are dependent on the green plants for food. There are many remarkable similarities in the metabolic pathways of various organisms and the manner in which we all derive our energy from the food we consume. In fact, much of what we know about human nutrition and biochemistry was worked out with experiments involving organisms as simple as bacteria or as complex as rats and mice.

ANSWERS TO QUESTIONS

1. (1) two carbon (2) ATP (3) nitrogen balance (4) positive (5) negative

2. Incorporated into protein, utilized in other syntheses, deaminated to keto acid.

3. (1) phenylalanine + H_2O + O_2 ⟶ ⌬—CH_2—C(=O)—COOH + NH_3 + H_2O

   (2) glycine + HOOC—CH_2CH_2—C(=O)—COOH ⟶

   HC(=O)—COOH + HOOC—CH_2CH_2—CH(NH$_2$)—COOH

4. (1) O (2) L (3) L (4) L (5) O (6) L (7) O

5. (1) 9 (2) 9 (3) 7 (4) 5 (5) 10 (6) 3

6. (1) d. Embden-Meyerhof
    (2) a. substrate level phosphorylation
    (3) e. citric acid cycle
    (4) b. electron transport-oxidative phosphorylation
    (5) c. photosynthesis

7. (1) arginine
    (2) a carboxylic acid

8. Lipogenesis–

    Reduction of keto group on β-carbon to form alcohol (hydrogenation)

    Dehydration between α and β carbons to form C = C

    Reduction of C = C between α and β carbons to form alkane (hydrogenation)

    Result–chain lengthened by two carbons

    Fatty acid oxidation–

    Oxidation of alcohol on β-carbon to form ketone (dehydrogenation)

    Hydration of C = C between α and β carbons

    Oxidation of α, β carbons to form double bond (dehydrogenation)

    Result–chain shortened by two carbons

9. Glucogenic amino acids produce glucose or glycogen. Ketogenic amino acids produce acetyl-CoA.

10. Activation, oxidation, hydration, oxidation, cleavage.

11. L-glutamic acid is transformed into L-glutamine by forming an amide bond. The L-glutamine transfers the nitrogen.

12. (1) (ii)
    (2) Most naturally occurring fatty acids contain an even number of carbons
    (3) 18

13. Products are:

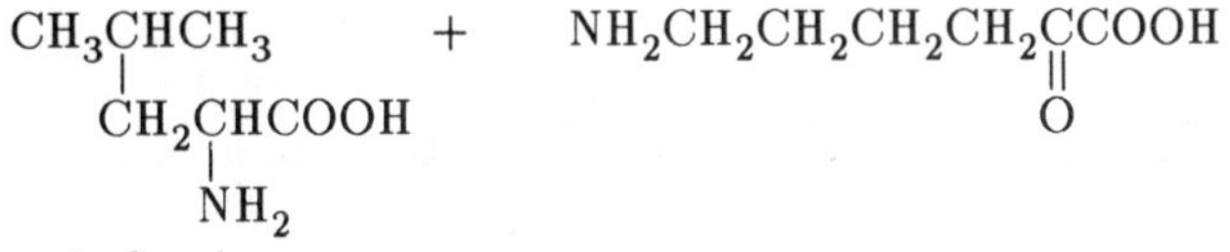

    CH_3CHCH_3 + $NH_2CH_2CH_2CH_2CH_2CCOOH$
    $|$ $\|$
    $CH_2CHCOOH$ O
    $|$
    NH_2
    isoleucine

14. amino acids away from pool (3), (4), (5)
    amino acids to the pool (1), (2), (5), (6)
    two way processes (5)

15. (1) is not   (2) FADH$_2$ (3) NADH   (4) oxidative phosphorylation
    (5) aerobic   (6) oxygen